AF710500

WITHDRAWN

Soldering for Electronic Assemblies

MANUFACTURING ENGINEERING AND MATERIALS PROCESSING

A Series of Reference Books and Textbooks

SERIES EDITORS

Geoffrey Boothroyd

Chairman, Department of Industrial and Manufacturing Engineering
University of Rhode Island
Kingston, Rhode Island

George E. Dieter

Dean, College of Engineering
University of Maryland
College Park, Maryland

1. Computers in Manufacturing, *U. Rembold, M. Seth and J. S. Weinstein*
2. Cold Rolling of Steel, *William L. Roberts*
3. Strengthening of Ceramics: Treatments, Tests and Design Applications, *Henry P. Kirchner*
4. Metal Forming: The Application of Limit Analysis, *Betzalel Avitzur*
5. Improving Productivity by Classification, Coding, and Data Base Standardization: The Key to Maximizing CAD/CAM and Group Technology, *William F. Hyde*
6. Automatic Assembly, *Geoffrey Boothroyd, Corrado Poli, and Laurence E. Murch*
7. Manufacturing Engineering Processes, *Leo Alting*
8. Modern Ceramic Engineering: Properties, Processing, and Use in Design, *David W. Richerson*
9. Interface Technology for Computer-Controlled Manufacturing Processes, *Ulrich Rembold, Karl Armbruster, and Wolfgang Ülzmann*
10. Hot Rolling of Steel, *William L. Roberts*
11. Adhesives in Manufacturing, *edited by Gerald L. Schneberger*
12. Understanding the Manufacturing Process: Key to Successful CAD/CAM Implementation, *Joseph Harrington, Jr.*
13. Industrial Materials Science and Engineering, *edited by Lawrence E. Murr*
14. Lubricants and Lubrication in Metalworking Operations, *Elliot S. Nachtman and Serope Kalpakjian*
15. Manufacturing Engineering: An Introduction to the Basic Functions, *John P. Tanner*
16. Computer-Integrated Manufacturing Technology and Systems, *Ulrich Rembold, Christian Blume, and Ruediger Dillmann*
17. Connections in Electronic Assemblies, *Anthony J. Bilotta*

18. Automation for Press Feed Operations: Applications and Economics, *Edward Walker*
19. Nontraditional Manufacturing Processes, *Gary F. Benedict*
20. Programmable Controllers for Factory Automation, *David G. Johnson*
21. Printed Circuit Assembly Manufacturing, *Fred W. Kear*
22. Manufacturing High Technology Handbook, *edited by Donatas Tijunelis and Keith E. McKee*
23. Factory Information Systems: Design and Implementation for CIM Management and Control, *John Gaylord*
24. Flat Processing of Steel, *William L. Roberts*
25. Soldering for Electronic Assemblies, *Leo P. Lambert*

OTHER VOLUMES IN PREPARATION

Soldering for Electronic Assemblies

Leo P. Lambert
Digital Equipment Corporation
Tewksbury, Massachusetts

Marcel Dekker, Inc. New York • Basel

TT
267
.L33
1988

Library of Congress Cataloging-in-Publication Data

Lambert, Leo P.

Soldering for electronic assemblies.

(Manufacturing engineering and materials processing ; 25)
Includes bibliographies and index.
1. Solder and soldering. 2. Printed circuits. I. Title. II. Series.
TT267.L33 1987 621.381 87-22371
ISBN 0-8247-7681-X

Copyright © 1988 by MARCEL DEKKER, INC.
All Rights Reserved

Neither this book nor any part may be reproduced or transmitted in any form or by any means, electronic or mechanical, including photocopying, microfilming, and recording, or by any information storage and retrieval system, without permission in writing from the publisher.

MARCEL DEKKER, INC.
270 Madison Avenue, New York, New York 10016

Current printing (last digit):
10 9 8 7 6 5 4 3 2 1

PRINTED IN THE UNITED STATES OF AMERICA

To Ellen, Anne, and Kate

for their support and understanding

Preface

The purpose of this book is to outline the specifics of soldering as it is performed by the electronics industry. The entire spectrum of electronic soldering is covered including the preparations required to perform quality work. The book discusses basic soldering processes, such as manual and automatic mass soldering techniques, cleaning processes, and surface mount process development encountered during the printed wiring assembly manufacturing cycles.

The need for more densely populated printed wiring assemblies is driven by the new technological advances in the areas of component interconnection. This increase in density creates many changes in component configurations and material selections. Traditional solderability testing procedures must be reviewed and modified to accommodate the new component types, and material selection must be critically analyzed to qualify the electrical and mechanical requirements of such devices and the assemblies.

This book outlines and defines these steps as logically as possible, so that after it is reviewed and studied, a soldering process can be justified, implemented, and made to function within a manufacturing process with minimum effort. It is written for the people who are directly involved with the soldering operations. Design engineering personnel will be able to review and develop an understanding of the processes during the product design stages; with this understanding, product can be designed to meet the requirements of manufacturing productivity and producibility. Those in support engineering functions also will be able to review the manufacturing steps and develop complementary processes to improve productivity. This necessitates the review of incoming inspection criteria and component specifications for meeting the solderability requirements.

The information in this book comes from 20 years of experience in the soldering and manufacturing processes. Additional information has been derived from attending many technical symposia and reviewing technical publications.

I would like to thank everyone, especially my co-workers, without whose aid this book could not have been written. I also wish to express a special thank you to Digital Equipment Corporation for all the support given me in this undertaking.

Leo P. Lambert

Contents

Soldering for Electronic Assemblies

1

Introduction

1.0 INTRODUCTION

The purpose of this book is to fill a void which I believe exists in the literature of soldering electronic assemblies. Many texts have been written describing the theoretical aspects of metallurgy, how to design solder joints, chemistries and their development, etc., but none which have actually described how to perform the soldering operation, or show how to justify the equipment and treat the process as a system. This text is intended to fill that gap.
It has been written to help engineers, supervisors, and operators understand how to establish process parameters. Each chapter is written to stand by itself. As an example, regarding fluxes, types and usage which apply to the process are discussed in the chapter with the process, with the intent of showing the intimacy of the

flux and the soldering process. Similarly, the same applies to the solder and its elements relative to the soldering process.

1.1 SEQUENCE FOR PROCESS FLOW

Chapter sequence follows traditional process flow, from product design through to the assembly and soldering phases. Methods and ideas are also given to augment the knowledge and skills of the manufacturing support personnel.

To understand everything about printed wiring assembly, including the elemental board and component fabrication process, would be a cumbersome task. The uniqueness of each element requires its own specialists and experts. In effect, each piece is further subdivided into smaller groups with its own experts and knowledge engineers. As can be seen, this further complicates the process and no one individual would have enough time and energy to maintain the knowledge successfully. There is just too much to learn in detail to be proficient in all the design and fabrication processes. Additionally, the academic disciplines vary for each process, such as chemist and chemical engineering for board fabrication, electrical engineering for product design, mechanical engineering for product packaging, industrial engineering for process flow, and manufacturing engineering for assembly. Professionals in all these disciplines have a distinct and vital part in the successful implementation of any product and process.

Therefore, although in-depth education of all disciplines is difficult to achieve, surface knowledge of each subject is necessary to assure comprehension of the process. This text will furnish you with the basics for asking pertinent questions on the subject matter and over a period of time increase hour proficiency to a level where you will be intimately involved with all features of the process.

The obstacle in developing this proficiency is the time required for participation in both formal and on-the-job training programs. Engineers look toward moving up the corporate ladder at an inflated or accelerated pace which can develop a false sense of security for the engineer and create an ineptness in the engineering staff of any corporation.

Engineering is a career, not a job; therefore continuous learning is a must, not a want or a would-like. To be a valuable contributor, one must have all the facts and analytical ability to develop and determine a conclusive response and recommendation to the problem at hand.

Reiterating process flow, the chapters are sequenced from the steps of designing the process from the product information phase through the manufacturing steps of assembling and soldering the boards. Emphasis on the soldering process for printed wiring boards (pwbs) is the main focus of the text.

Soldering is the most important process in the manufacturing cycle. All pwbs must be soldered by either manual or mass soldering techniques. When first establishing assembly processes, components and printed wiring boards are purchased. Then methods are designed and developed to test the components for conformance to requirements. Initially these test methods only checked for functionality of the product. Components must be checked for both electrical and mechanical requirements which includes solderability testing. Boards must also be inspected for both electrical and mechanical requirements including solderability.

1.2 CHAPTER ABSTRACTS

Chapter 2 covers the ingredients of soldering such as the selection of the fluxes, the soldering materials, and the cored solder used in the manual soldering operation. Each of these elements has a bearing on the final product and must be cautiously selected to produce a quality product.

Chapter 3 deals with the design of the process for the product being manufactured. Its theme is to generate questions to develop the knowledge of the product in order to make one proficient enough to determine how the product is going to be manufactured. Issues ranging from the electrical requirements to the quantity to be manufactured must be addressed to certify quality products from quality processes.

Hence, Chaps. 2 and 3 cover the steps preparatory to designing the process. They again force the engineer to learn about the product before designing the process. In this way the product will not be destroyed by the process.

Chapter 4 delves into soldering systems including the functions of each section as well as the operational procedures for mass soldering systems. It also discusses some methods of economically justifying equipment and developing process capacity models.

Chapter 5 takes an in-depth view of the cleaning processes. Methods of cleaning, be they solvent or water, are discussed including the equipment to accomplish such cleaning functions. It also touches on the safety requirements necessary to certify compliance to the environmental regulations established by the local regulatory agencies and the federal government. It does not state the requirements per se, but it does raise the questions to properly address the subject.

Chapter 6 reviews manual soldering. Over the years this process has been abused and left as an afterthought, when in fact it has created more problems than it has solved due to uneducated personnel performing the operation. Product changes and equipment enhancements necessitate the need for new training programs which will show how to choose and use the proper tools. Chapter 6 also covers the manual soldering theory for removing and replacing components in printed wiring assemblies. Repair of printed wiring assemblies is the most difficult soldering function to master and typically the operation which is given the least attention.

Chapter 7 introduces the reader to the surface mount process. Surface mount is being rejuvenated for component mounting due to advantages of space savings and improved electrical functionality. Although the majority of the industry is using these two reasons for converting from through-hole technology to surface mount, there are many elements which must be considered for successful conversion from one process to another. Chapter 7 identifies some of those concerns and asks questions to provoke further investigation for successful implementation.

Chapter 8 discusses troubleshooting and overall recommendations.

2

The Elements of Soldering

2.0 The Elements of Soldering
2.1 Fluxes
2.1.1 Flux Functions
2.1.2 Flux Requirements
2.1.3 Flux Types
Rosin
Rosin Mildly Activated
Rosin Activated
2.1.4 Rosin Characteristics
2.1.5 Organic Acid Fluxes
2.1.6 Selection of Fluxes
Product Application to Activity Levels
Flux Application to the Product
Example of Relating Flux Requirements to the Process
2.1.7 Tests for Flux Performance
Globule Test
Solder Spread Test
2.1.8 The Process of Choosing a Flux
Define the Product
Question the Expected Solderability Results of Components and Boards

2.0 THE ELEMENTS OF SOLDERING

For proper soldering to occur, many factors and elements must be considered, not least of which are the chemicals and materials. Chemicals include various types of fluxes used in the removal of metallic oxides or the reactionary product of oxygen and the base materials being soldered. Additionally, they must prevent these reactionary products from recurring during the soldering operation. Hence, in designing these chemicals for soldering, consideration must be given to the surface plating of both the boards and the devices to be soldered.

In this chapter, the chemicals and materials necessary to solder will be discussed in detail.

2.1 FLUXES

By definition, the purpose of flux is to remove metallic oxides and tarnish films to prepare the metallic surfaces for soldering. Preparing metallic surfaces of soldering includes not only the removal of oxides or reactionary oxygenated byproducts but also the protection of those surfaces during the soldering operations. This implies that flux will prevent reoxidation of the metallic surfaces and promote wetting of the solder. The development of formulation of fluxes are very complex mechanisms and will not be discussed in this text. Again, a purpose of this text is to help the engineer and support groups in the flux selection process and identify the points of caution necessary to prevent catastrophic failures of any electrical product.

The following flux categories are listed from the lowest activity to the highest. The higher activity levels signify higher percentages of activators to attack the metallic oxides. These activators can be either halogens or organic acids.

- Rosin
- Rosin mildly activated
- Rosin activated
- Rosin superactivated
- Synthetic rosin
- Organic acid
- Organic acid halogen-free
- Neutral organic acid
- Inorganic acid

See Table 2.1.

2.1.1 Flux Functions

Fluxes perform three functions in the soldering operation:

Table 2.1 Flux Activity Chart

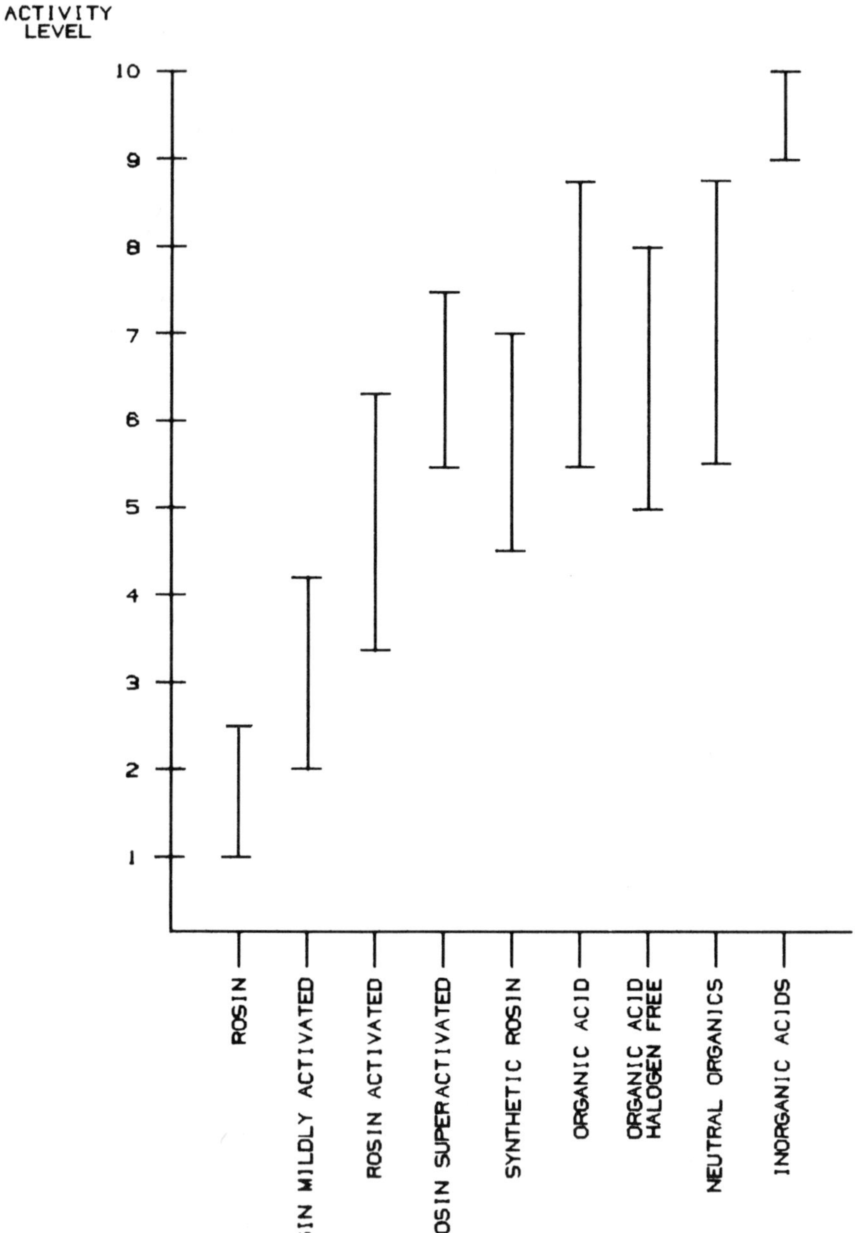

- Chemically: fluxes remove metallic oxides and tarnishes.
- Thermally: fluxes enhance the heat transfer from the molten solder to the joint or metals being joined during the soldering operation.
- Physically: fluxes are used to promote the wetting, i.e., the metallurgical attraction of a liquid metal to solid metal. Wetting is the ability of the solder to spread evenly over the surfaces that have been prepared by the flux. A properly wetted surface will have a smooth, clean, unbroken film of solder over the entire surface of the base materials [1].

2.1.2 Flux Requirements

In choosing a flux, the following requirements should be considered:

Flux must:

1. Remove oxide and penetrate films.
2. Prevent oxidation at soldering temperatures.
3. Lower interfacial surface tension.
4. Be easily displaced by the molten solder.
5. Be harmless to components.
6. Be easily removed if desired.
7. Be thermally stable [2].

2.1.3 Flux Types

Rosin

Rosin fluxes (type R) are a combination of rosin (abietic acid) and plasticizers and are nonchlorinated. Chlorinated materials would change the characteristics of the flux, increasing its activity and aggressiveness. Rosin flux is weak and used only when

- o Excellent soldering conditions exist.
- o Postsolder cleaning is prohibited
- o Corrosive (ionic) residues cannot be permitted to remain on the board.

Rosins come from pine tree sap and are graded by color. "The purest form of rosin by the ASTM designation is called 'Water White' or WW for short" [3]. Military specification MIL-P-14256 identifies rosin flux as type R and states how the flux shall be composed and the specifications required for its formulation.

Rosin Mildly Activated

Mildly activated rosin flux (type RMA) contains low levels of nonchlorinated or neutral activators to increase the activity levels during soldering temperatures. Neutral activators are defined as activators needing heat to begin the chemical reaction. At room temperature, RMA flux activators are not chemically active and will not attack the metallic oxides. This is continuously changing with material improvements and should be checked carefully to certify the activity levels are in fact of the RMA flux category.

MIL-P-14256 states that when fluxes are tested, the arithmetic mean of the specific resistivities of the water extracts shall be at least 100,000 ohm-cm.

Water extract tests are used to measure the resistivity or, conversely, the conductivity of flux activators. This is accomplished by fabricating a control sample of distilled water with a resistivity reading of 500,000 ohm-cm. With a portion of this distilled water sample, the flux to be tested is added in a controlled amount and boiled for a given time period, usually 1 min. Once boiled the sample is allowed to cool for about 10 min then it is placed in a bath of water until it reaches 23°C ± 2°C (73.4 + 3.6°F). At this time, resistivity measurements are taken of both the water control sample and flux water extract sample. The water control sample must always be above 500,000 ohm-cm at the control temperature. It it isn't, the test must be repeated. Detailed information regarding the exact procedures can be found in the military specification or the EIA (Electronic Industries Association) Standards RS-402.

The activators in RMA fluxes are typically nonactive at ambient temperature but become active at soldering temperature. This is the main reason some specifications allow RMA fluxes to be left behind uncleaned. At working temperatures the fear of insulation resistance breakdown by reaction of moisture with the RMA flux activators is very low.

Rosin Activated

Activated rosin fluxes (RA) have the highest level of activators of the rosin family. The activators used are usually amine salts of chlorides or bromides. These activators are active at room temperature as compared to the activators used in RMA which are activated by the heat of the soldering operation.

Federal specification QQ-S-571 requires RA fluxes to have a water extract test of 45,000 ohm-cm. There are many formulations of RA fluxes to accommodate both industrial and military applications. Industrial RA fluxes usually have a higher level of activation and are sometimes referred to as superactivated rosin fluxes (type SRA).

2.1.4 Rosin Characteristics

Although we have only discussed the activity levels of the rosin types of fluxes, the rosin content also varies within each level-type flux. Rosin starts to melt at 160°F (70.4°C) and reacts with copper at approximately 225°F (106°C). At 500 °F (257.4°C), the typical soldering temperature for electronic assemblies, fluxes begin to decompose, causing their greatest activity. At higher temperature, 650°F (340°C) or above, as experienced during hand soldering, the flux polymerizes and turns into a crystalline mechanism that is very difficult to remove with any type of solvent [4].

To increase the activity level of (RA) fluxes, various chloride or bromide salt activators are added. In the lower activated fluxes, such as type (R) rosin or (RMA) rosin mildly activated, the activators used are not active at room or ambient temperature. As previously stated, these RMA activators need heat to be activated. However, the activators in RA rosin activated fluxes are active at all temperatures and must be closely monitored because as long as they are in contact with the base material, a reactionary chemical function is occurring. The fluxes etch away the metallic oxides or, if they have been completely removed, attack the base material.

Therefore, to get the proper fluxing action of low-activity rosin fluxes such as types R and RMA, the rosin solids contents must be kept high, such as 35-45%. Rosin itself is a form of abeitic acid, and therefore it also has corrosive properties. One way to get more activity out of the rosin is to increase the rosin content of the flux. Fluxes are a solution of alcohols and rosin, and the solids percentage indicates the amount of rosin in the flux. Synthetic fluxes or modified rosin called resin fluxes may not have that particular capability and may need to be of higher organic activators to be of comparable activity.

The term *resin* is used to describe a manufactured product. Rosin is a natural product harvested from pine trees and fluxes using rosin have not been modified. Resin is a chemically fabricated rosin, or synthetic rosin. When the resin is combined with the alcohols and various other raw ingredients of the flux system, the entire product is then identified as a resin flux.

Industrial fluxes usually have more activators added and use a lower percentage of solids. This low solids content enhances the foaming ability of the flux. Foaming the flux is an application method used in mass wave-soldering systems. High solid content fluxes are very poor foamers and are typically applied with a wave fluxer system. If a foam fluxer is used, then the ability of the flux to foam is a necessity. As will be discussed later, when a board goes through the foam and the foam breaks on the surface to be soldered, an even coat of flux is applied over all the surfaces, preparing the product to be soldered.

Low solids content fluxes are also easier to remove during the cleaning cycle. This is an important parameter of soldering electronic assemblies to prevent any electrical and corrosion degradation problems.

Chap. 5 will discuss removal of flux residues; however, it must be kept in mind that rosin fluxes consist of nonpolar materials with polar activators. Bipolar solvents are therefore required to remove the rosin and nonionic contaminants. It must be noted that ionic residues left on the board will cause electrical surface insulation breakdown if exposed to moisture.

2.1.5 Organic Acid Fluxes

Another class of flux is the organic acids (OA), potentially misnamed due to the fear they impose on the users. This fear is based on the term *acid*; whenever acid and metal surfaces are brought together the possibility exists that the acids will create some reactionary process with the metal and cause a contamination problem which will impact the electrical functionality of the product.

Another name for acid fluxes is "water-soluble fluxes." Originally, water-soluble fluxes had water as the base ingredient but water caused problems during the soldering process. The water caused the solder to spatter and create many solder balls, voids, and other solder-related problems. In recent years, water as a base ingredient has been replaced by alcohol, thus eliminating the problem.

The pH levels of organic acid fluxes are as low as 2.0. The pH levels are used for comparative purposes in the evaluation of various fluxes. Usually they are composed and adjusted with the following: alcohols, organic acids, and halogenated salts or other acid types. Some have ethanolamines, alkylene glycols, polyglycol, and other types of similar materials [5].

Factors to consider when choosing organic acid fluxes are the requirements of the modules as dictated by the design of the product, e.g., component height off the board, density of components, and the contractors' processes. In addition, the ability of the flux to promote wetting and remain soluble enough to be cleaned off the printed wiring assembly and from beneath components must be reviewed for all product types to be soldered. There must be no entrapment areas on the modules when using water-soluble flux. The encroachment of via holes beneath any flat-bodied component is a potential source of contamination. If via holes exist beneath components, the flux will be in place after the fluxing step and sealed when the solder fills the via while going over the wave. When this occurs, removing the residues is difficult with any solvent type, including water; therefore components should be designed to be assembled off the surface of the board. Components design and requirements should also be identified in the purchase specifications.

2.1.6 Selection of Fluxes

Although many types of flux exist, the process of deciding which one to use can become very complicated depending on product specifications and design requirements.

Product Application to Activity Levels

Typical commercial applications require product functionality and reliability for a given time span, ranging from 6 months to 5 years. Industrial applications usually require 5-7 years, whereas the telecommunications industry requires 30-40 years and military products depend on specific use.

Specifications usually identify the fluxes to be used for various products. Military applications require the use of RMA fluxes, although RA fluxes are at times allowed. RMA fluxes are required due to their nonactive nature at room temperature and when certain circumstances occur where they are entrapped beneath components they will not be a threat to product reliability. However, the burden of proving that the flux residues can be removed from the assembly rests with the manufacturer of the product. Commercial specifications are not as stringent as the military, and therefore stronger fluxes may be used in their soldering process.

The Institute for Interconnecting and Packaging Electronic Circuits (IPC) identifies in its specification IPS-S-815 three categories or general classes for all electronic assemblies:

Class 1. Consumer product. Includes TV sets, toys, entertainment electronics, and noncritical consumer or industrial control devices.

Class 2. General industrial. Includes computers, telecommunication equipment, sophisticated business machines, instruments, and certain noncritical military applications.

Class 3. High reliability. Includes equipments were continuous performance is critical, equipment downtime cannot be tolerated, or equipment is a life support system.

And the IPC includes the statement: "The user has the responsibility to determine the class into which the electronic interconnections belong" [6].

As can be seen, the soldering results to be accomplished are only part of the total criteria for choosing a flux. Other requirements, as previously stated, must also be considered and weighed.

Flux Application to the Product

Flux application methods must be considered when specifying flux types. There are many methods utilized; however, the most prevalent are foam- and wave-type fluxers. Spray fluxers are also effective in certain cases.

Applying flux to the board and component surfaces is not simply a matter of passing a board through the materials and then conveying it over the solder wave. The quantity and quality of flux used must be monitored continuously throughout the operation.

The quantity of flux in each type fluxer pot may vary, but for each system the proper level must be established and maintained. If the flux level gets low, then improper dispensing will occur and poor fluxing will result.

After the application of flux, the excess must be removed either by wiping away the large droplets with wipeoff brushes or activating a low-pressure air knife to squeegee the flux off the bottom of the board. Removal of excess flux is necessary to prevent the droplets from dropping into and onto the preheaters. Any flux accumulation dripping into this area is dangerous and could cause a fire. We will discuss this in more detail in Chap. 5.

The quality of flux is tremendously important and cannot be overemphasized. The specific gravity must be measured at least twice per day and preferably every 2 hrs. The flux specific gravity cannot be allowed to vary because any variation in specific gravity will affect the soldering process by applying either too little or too much material. Therefore controls are necessary!

Quality of flux also includes contamination within the flux, such as oils, water, and silicones. The operation must be kept clean. Soldering systems with finger conveyors must have finger cleaners and modifications to keep the returning fingers from passing through the flux.

The following flux questions must be answered: Will it flux all the surfaces and prepare them for soldering? Can the flux be cleaned from the board's surfaces if it is necessary due to product requirements? A "yes" is required for both questions.

As previously stated, some fluxes can remain on the board whereas others have to be removed. In any case, it is good practice not to leave any activated fluxes on the board. Certain fluxes are being designed to be left on the board and their residues are supposedly inert and the film is hard and nonconductive. To use this type of flux, an understanding of all aspects of the end results must be known to certify nondetrimental short- and long-term effects on the product.

Example of Relating Flux Requirements to the Process

When conducting flux experiments to determine flux variations and activity levels that would be experienced with the flux types R, RMA, RA, SRA, and OA, all the process parameters must be held constant except for the flux itself. Doing this can provide a quantitative comparison of the effectiveness of the tested fluxes.

Some experimental rules of thumb are as follows:

- Solder pot should use a lower solder temperature than the production solder system. Test pot = 460°F (235.4°C); Production solder pot = 500°F (257.4°C).

- Immersion rate should be as specified by the testing equipment.
- Dwell time in solder pot should be 3-4 sec longer than would be used on the production machine.

Example: Soldering time on a mass solder system is typically 1 to 1-1½ sec, using 5 feet per minute (FPM) conveyor speed. Therefore, to test the flux properly, the test time should be approximately 5 sec. This is to provide an overall picture of the wetting curves to quantify the flux properly. However, in acceptable conditions the test sample or coupon will solder within 2 sec. If it takes any longer, excessive heat might ruin the laminate material.

2.1.7 Tests for Flux Performance

To measure the activity levels of the flux, a few tests are available, such as:

- Copper mirror
- Chloride extract
- Resistivity of solution

Military specification identify the methods of conducting the tests and interpreting the results. This text will not discuss these tests, as there is information available in the public domain, describing methods, procedures, and acceptable results. Still, it is important to perform these tests to certify the activity levels of fluxes. Changes in activity levels can have dramatic effects on the soldering process.

Globule Test

As mentioned in Chap. 3, the globule test is utilized to measure the solderability of axial leaded component leads. However, the test can also be used to measure the effectiveness of fluxes. The procedure for the evaluation is holding all the elements constant and making the flux the variable. In this way, different fluxes can be compared for their attributes in relation to the actual soldering times.

Modifications of this test are also used for measuring solder "rise time" in plated through holes. The concept is similar to the standard globule test for component leads, except the plated through hole is placed over the molten solder pellet and the time the solder takes to rise in the plated through hole is measured. This is an important test for printed wiring boards because it is a nondestructive test and does not destroy the module, unlike the wave solder test. The holes used for the solder rise test are via holes, or interfacial holes, leaving the component holes open for future component insertion (see Fig. 2.1).

Solder Spread Test

The solder spread test identifies the ability of any flux to promote lateral wetting on a copper coupon. The test consists of placing a copper coupon on a hot plate, applying a given quantity of flux, positioning a solder pellet onto the heated sample, allowing the solder pellet to melt and spread over the surface of the coupon. The area that the solder has spread over the coupon is measured. The larger the spread, the better the wetting ability of the material being tested.

The solder spread test is used to measure the effectiveness of the fluxes in preparing a surface for the acceptance of solder. A suggested procedure is:

1. Measure the thickness of the sample coupon.
2. Apply flux to the coupon. The quantity of flux must be consistent from coupon to coupon.
3. Position solder pellet onto coupon.
4. Position coupon on hot plate, preheated to 495 ± 5°F (254.6°C).
5. After the solder pellet melts, allow it to stay in place for a given time period.

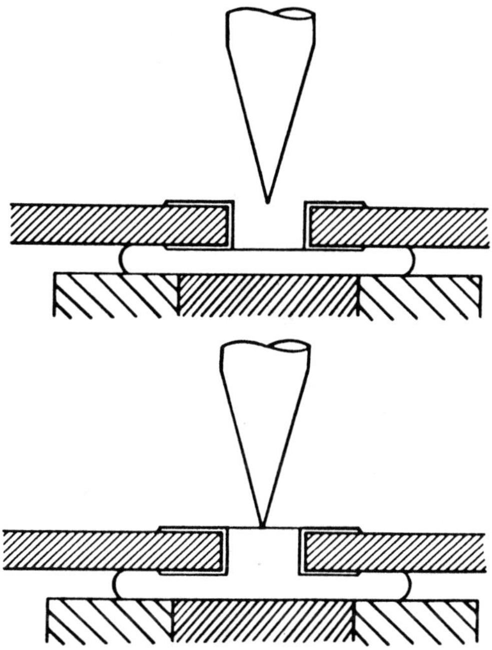

Figure 2.1 Plated through hole time rise test. Courtesy of Multicore Solders Ltd.

6. Gently remove coupon from hot plate and allow solder to solidify.
7. Clean coupon in appropriate solvent.
8. Measure overall thickness of coupon and subtract the original coupon thickness identified in step 1.
9. The difference of the thickness relates to the spread of the solder. Comparisons can then be made identifying various extents of spreading.

2.1.8 The Process of Choosing a Flux

In choosing a flux, the following must be considered:

- Will it enhance solderability? Comparisons should be conducted with a variety of similar fluxes to determine which one or which groups work the best for your product.
- Can it be easily removed to prevent electrical degradation of the product?
- Will it enhance the reliability of the product?

As earlier stated, choosing a flux for soldering electronic assemblies is a process that must be monitored closely. Once the flux has been defined, the cleaning process must be designed for removal of such flux, and this is dependent on the ingredients of the flux. The process recommended by the author is to ask the following questions and answer them in relation to the product being manufactured.

Define the Product

- Is the end product military, commercial electronic, or consumer electronic?
- What is the expected product life?
- What are the electrical requirements?
- Define the resistance properties of the components and circuitry.
- Define the metal coatings and the thickness to be soldered.
- Identify the solderability requirements.

By defining the product and its expected reliability performance, the proper process and quality requirements can be developed and implemented. This includes defining the criteria for purchasing of the correct components and laminate materials.

Question the Expected Solderability Results of Components and Boards

- What are the solderability requirements?
- Are the components tested for solderability?
- Do the components and boards pass the solderability requirements?
- Are the plating and plating thickness specified in the component specifications?
- Is storage handled on a first in/first out (FIFO) basis?
- Is any metallurgical work performed on lead plating, lead thickness, and intermetallic structures to improve solderability on parts to be soldered?

Question the Compatibility of the Board Design

- Is the design of the board compatible with the process?
- Are the components mounted flush to the board, or are they mounted off the surface of the board, to allow the cleaning solvent to penetrate beneath the component for proper cleaning action?

Question the Fluxer Type

Questioning the type of fluxer used forces the research necessary to properly define the best fluxer mechanism for the chemistries or chemicals being used and the product being soldered. Hence asking how the flux is to be applied necessitates the review of the methods shown in Table 2.2.

Question the Cleaning Process

- What kind of cleaning operation is being used?

Table 2.2 Advantages and Disadvantages of Flux Types

	Advantages	Disadvantages
Foam	Ability to penetrate plated through holes Low solids contents Easily cleaned Ease of control	Specific gravity sensitive Product temperature sensitive
Spray	Consistency of application	Overspray Clogged nozzles High maintenance
Wave	High solid content Not sensitive to specific gravity Ease of wave height adjustment	Difficult to clean Poor penetration into product Large quantity of flux applied to product
Dip	Minimal equipment cost	Inconsistent application Potential flux stratification

- How was the cleaning process determined? Aqueous, aqueous with detergent, or solvent? If solvent,
  - Bipolar
  - Chlorinated
  - Fluorinated
- Is the process a batch or in-line configuration?

Question Component Compatibility with Fluxes

- Are the components compatible with the fluxes?
- Are the components sealed to prevent the flux from attacking the body or seeping within the device itself?

- Is the marking on the components compatible to chemistries of the process?

Question the Residual Effects of Leaving Flux Behind

- What will happen if the flux is not removed immediately? Within 20 min, 60 min, or 1 day?
- Is the flux allowed to remain on the surface for a period of time? How long?
- Have all insulation resistance tests been conducted to determine the degradation of product when the flux is not removed?

Question the Type of Solder Mask

- Is solder mask used on the board?
- Does the board have permanent solder mask or is it just laminate material?
- Does it lift off, called mealing, the board due to poor adhesion to the laminate and allow the flux to seep between the mask and the laminate?

These questions must be addressed by the disciplines of design engineering, manufacturing engineering, and production engineering, all responsible for the process, to certify that no detrimental effects will result from the flux chemistries used during the soldering process.

2.1.9 Test for Measuring Effectiveness of Flux Removal Process

Some of the tests necessary to certify proper flux removal are:

- Black light
- Solvent extract

- o Insulation resistance
- o Temperature humidity cycling

These tests define whether or not any residual components of the flux remain on the surface of the modules. More information on cleaning is given in Chap. 5.

2.2 SOLDERING

Soft soldering is performed between the temperatures of 160°F (70.4°C) and 800°F (422.4°C). Its function is to use a distinct metallic compound such as tin-lead to join two of the same or different metals.

2.3 SOLDERING MATERIALS

2.3.1 Solder Alloy

"Alloy is a mixture of metals made to produce desired characteristics" [7]. Alloys are created by combining two or more metals while they are liquid and allowing them to solidify. The alloying process produces a new material that will be different from each of its constituents. Varying the percentages of each metal in the alloy will create different characteristics in the new product.

Lower melting alloys are created by experimentation with various percentages of metallic ingredients. Figure 2.2 shows the melting point of tin at 450°F (232°C) and lead at 621°F (327°C). By alloying the tin-lead to a 63-37% solution, the melting point is lowered to 361°F (183°C), the eutectic point of the tin-lead alloy composition. At the eutectic point the heat of fusion occurs and the the material goes directly from solid to liquid.

The eutectic point of any metal is the point where the material goes immediately from a solid phase into a liquid phase without going through a transitory plastic range. As can be seen on the phase diagram (Fig. 2.2), the eutectic point is at 63% tine and 37% lead.

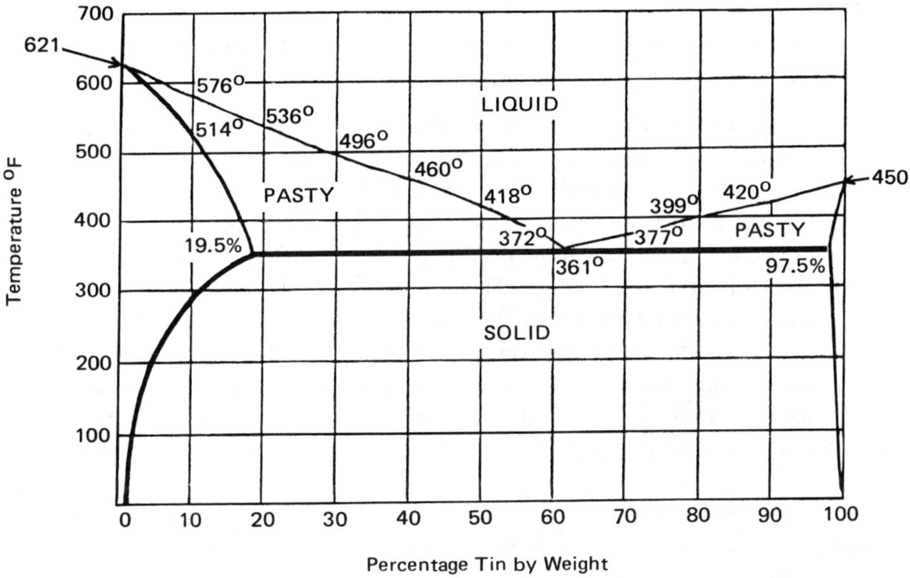

Figure 2.2 Tin lead phase diagram. Adapted from the IPC Assembly and Joining Techniques Number 4.0.1.

Solder is composed of two major ingredients: tin and lead. Other ingredients such as bismuth or indium are added to perform functions such as lowering the eutectic point; silver is added to prevent silver migration; and antimony is added to prevent or reduce tin pest. Tin pest is the decomposition of tin into a powdery substance when pure tin is exposed to cold temperatures.

Both tin and lead are taken to refineries for processing, where tin is reduced to the molten state, and the impurities are removed, rendering it relatively free of other metallic and nonmetallic contamination. It is then alloyed with the lead to the percentages defined by the product mix. If the solder is produced with tin from the ore mines, then the solder is identified as virgin grade solder. If it is reclaimed from an old solder operation, then it is identified as electronically refined solder.

2.3.2 Tin

> Tin is obtained from deposits of the mineral ore Cassiterite. The ore is found in veins in rock and gravel. . .
>
> How Pure Tin is Made. After the Cassiterite leaves the mines, it goes through five important processes before pure tin is obtained. Cassiterite contains arsenic and sulfur. These materials are burned out of the ore in a furnace or a roasting machine. When copper sulfide is also contained in the ore, it is drawn off by a process called leaching. The ore then is heated in a furnace. This process is called smelting. The ore is heated for about six hours. The smelting causes the impurities, or slag, which are lighter than the tin, to rise to the top. The slag is drawn off, and the metal is heated for about six hours more. The tin, in liquid form, sinks to the bottom of the furnace and is drawn off and runs into molds to cool as ingots [8].

The largest tin mines are found in Malaysia, Bolivia, and the Netherlands. The United States produces very little tin for the commercial market.

2.3.3 Lead

> Lead is found in veins and masses in Limestone and Dolomite. It is usually found with deposits of other metals such as Zinc, Silver, Copper, and Gold. When Lead is the only metal found in a rock deposit, the Lead content must be more than 5 percent by weight in order to make it practical to mine the metal. The most common Lead-ore mineral is Galena, a combination of Lead and Sulfur which is called Simple Sulfide of Lead. Another ore mineral in which Lead is found combined with Sulfur is Anglesite. Cerussite is a Lead ore mineral which contains carbon [9].
>
> The ore is broken from the deposit and brought to the surface to be crushed and refined, or separated from the other minerals in the ore in a process called Concentration. The two chief ways of concentrating Lead are smelting and floatation [10].

The major countries supplying lead to the world are the United States, Canada, Burma, and Germany.

2.3.4 Refining Reclaimed Solder

Refining is used to reclaim old or used solder and is another method of generating electronic grade solder. These processes were developed to meet the specification of QQ-S-571, due to the problems of importing the raw materials into the United States during the Second World War.

Some impurities are difficult and expensive to remove and, depending on the future uses of the solder, may not be removed. Sulfur is a typical element of this category. It is used in removing contaminants from molten solder and, if not properly handled, remains behind in the main product. Sulfur causes a problem when it intermixes with solder; a large quantity of dross (tin-lead oxide) is produced. This dross contaminates the solder joints, being drawn into the solder machine's pumping mechanism and intermixing in the molten solder.

Previous metals and noble metals are always removed for their intrinsic value, while copper is removed to adjust the alloy content to meet QQ-S-571.

One base material of a printed wiring board is copper, and copper dissolves in solder. This phenomenon removes tin from the solder pot and causes tin depletion. Therefore the level of tin must always be monitored during soldering to ensure a correct soldering alloy. Monitoring the tin is done in a lab with either atomic absorption techniques or spectrographic analysis.

2.3.5 Selecting the Best Alloy for the Process

The most widely used alloys in electronic soldering are 63-37 and 60-40 tin-lead, the reasons being the eutectic characteristics of the 63-37 alloys.

The eutectic alloy 63-37 transforms from solid to liquid at 361°F (183°C), and the major advantage is the immediate solidification at this temperature during cooling. Rapid cooling prevents movement of the solder and components during solidification. Slow solidification can cause "disturbed solder joints" which are characterized by solidification lines on the surface of the joint.

Electrically environmental testing and thermal cycling of disturbed solder joints could become intermittent and create catastrophic failures.

Solder Pricing

The price of solder is based on the prices of the base materials tin and lead. The price of virgin tin on the metal market should always be known by the buyer when determining whether a quoted price is reasonable.

Example: Price of tin =
Price of lead =

For 63-37 solder, the price equals 63% of tin market price plus 37% of lead market price, plus the manufacturing and handling charges as defined by the manufacturer. These prices are available from the daily metals market in the business section of the newspapers.

With this information one can predict with reasonable accuracy the price of solder. Caution must be exercised, however, when buying refined solder, since this material is not controlled at market prices. Refined solder is made from used material and the base material is purchased by the vendors at a lower price than the market price and they should offer considerable discounts to the user. This is an important criterion when buying electronic grade solder from the refiners.

As can be seen then, a 60-40 alloy would be less expensive to use than a 63-37 alloy and the minor difference in the tin content then makes these two alloys useful for the soldering process. Although theoretically there is a melting temperature delta of 15°F between these two alloys, the percentage tolerance of tin between the two when measured against the standard specifications is only 1%. This is due to the tolerance buildup of the other metallic elements in the solder itself and the accuracies of the manufacturing process. In actuality, then, the 63-37 alloy is a theoretical number used to give credence to the fact that eutectic conditions will occur if the two elements of tin and lead are measured to the exact percentage. In actuality both are very similar and, although I would not recommend mixing the two alloys in a solder pot, both are individually acceptable for use in a mass soldering system.

In use, solder alloy should be monitored closely for elemental characteristics of all contaminants. Variations in the percentages can create situations whereby one type of solder is bought, i.e., 63-37, but due to tin depletion, for example, the tin level goes down into the range of 60-40 alloy and then behaves as a 60-40 alloy which is not eutectic. If the process has been designed to use 63-37 then monitoring of the solder contamination is necessary to keep the tin levels within the acceptable range. To be safe, the solder bath should be monitored monthly in order to prevent major changes from surprising the user.

2.4 CORED SOLDER

Initially developed for manual soldering, this form of solder has been available for many years. It is fabricated by extrusion of various dyes, and until the correct diameter is formed. It can be extruded either as a solid wire or with a core of flux. Plumbers solder, a 50-50% alloy, is usually solid wire solder.

Electronic grade wire solder is usually cored with a flux.

2.4.1 The Flux Core

The flux in the core is the same as the flux used in automatic soldering systems, except that it contains plasticizers to hold it in place at ambient temperatures (see Fig. 2.3).

2.4.2 Various Activity Levels and Their Uses

Data on various flux activity levels along with various flux percentages are available from the manufacturers.

Rosin flux cored solder is used only when soldering conditions are perfect, there are no soldering problems, and all surfaces are clean and not oxided. Also rosin cored solder does not have to be cleaned off the assembly when soldering is completed.

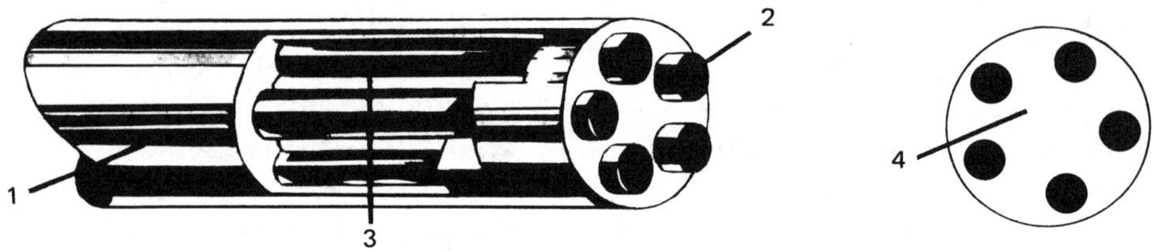

Figure 2.3 Multicore solder. This figure shows the construction of 5-core solder. Multicore solder is made only from pure virgin metals (1) with 5 cores of flux (2). Multicore's special manufacturing process locates the 5 cores of flux close to each other and at the perimeter, so that there is only an ultra-thin wall of solder between the cores of flux and the work (3). This unique core configuration (4) also keeps the flux near the heat source, to melt faster-wet the metal more effectively. This enables the solder to produce the consistently reliable joints. Courtesy of Multicore Solders Ltd.

The next higher activity level flux for cored solder is RMA, one of the most popular solders used by the electronics industry. In certain conditions this flux may not require removal. This is because its activators function only at elevated temperatures. At room tempratures they are nonactive and noncorrosive. Also, by being entrapped within the rosin, which is hydroscopic (rejects water or will not absorb water), the activators cannot absorb moisture from the environment.

Rosin activated (RA) cored solders are activated by an (RA) flux with activators quite different from those in an RMA flux. RA flux activators are usually organic acids or halogen activators and are active at all temperatures. Therefore any absorption of moisture will cause a reaction which could lead to dendritic growth between the component leads and the etch lines.

The activators are mixed within the rosin flux and since all the RA fluxes have different quantities and types of activators, it is important to check all materials for flux activity and corrosivity properties. Therefore, before endorsing any one material to be used in the process, tests should be conducted to verify the activity levels and the simulated environmental conditions of the product life cycle. This is necessary to develop a confidence level in the materials that are being used and to specify in specifications which fluxes are acceptable and which are not.

Water-soluble cored solders are finding wider use. However, when they are not properly specified, product failures occur. The activity levels within the fluxes are very high and the residues are not protected by anything, and will react with the moisture in the atmosphere. These residues must be cleaned off immediately after soldering.

Water-soluble flux by definition is soluble in water; however, the residues of water-soluble flux after the soldering operation are only soluble for a short period of time, approximately 15-20 min. Once dried, they convert to a metallic salt type of residue that is soluble neither in water nor in solvent. Hence, it is important to remove these residues immediately after soldering to prevent this conversion.

2.4.3 Problems with Cored Solder

The principle behind having flux in the core of the solder is to carry flux to the solder joint during the same motion or operation of bringing the solder to the joint.

Cored Solder Construction

The flux core is manufactured with plasticizers to keep the flux in a semirigid state within the solder wire. The plasticizers are designed to melt prior to the melting of the solder alloy, flowing down onto the surfaces to joints preparing them for soldering. Although this happens rather quickly, it is important to note that when it doesn't occur the solder will ball up and not wet the joint or parts to be soldered.

Without the flux core, the useful life of soldering tips will be shortened. Oxide will build up on the tip and will not wipe off during hand soldering. Also, flux voids often appear within cored solder. Such voids cause solder to ball up on the iron tip and prevent wetting.

The size of the core also has a dramatic effect on the creation of solder joints. Too low a percentage of core relative to metal will not supply enough flux to prepare the joint; too large a percentage will mean too little solder to create the joint. Three percent is the standard core for the electronics industry.

The correct procedure for determining the amount of cored solder to be used in filling plated through holes is as follows: measure the volume of both the PTH and the component lead, subtract the volume of the component lead from the volume of the PTH to determine the actual PTH volume. The next step is to determine the volume of cored solder that is equal to the plated through hole needing to be filled.

For a plated through hole with an 0.038 in. diameter finished hole, a good rule of thumb is no more than 3/16 in. of cored solder of 0.032 in. diameter. This is enough metal to allow for good fillets on both sides of the board.

Methods of Usage

A practice has been developed to prevent flux skip from creating nonwetted joints. The end of the flux wire is dipped into a cup of flux, creating a flux droplet on the end of the wire. This droplet is then carried to the solder joint and utilized in addition to the core to prepare he solder joint.

Styles of Cored Solder and Typical Applications

In manufacturing operations, water-soluble cored solder is used mainly in the touchup area, while rosin cored solders are used in final assembly, component removal, and replacement operations. The reason is the cleaning cycle utilized at various steps in the process. Touched-up printed wiring assemblies can be returned through the initial cleaning system, either solvent or aqueous. Parts assembled in final assembly are traditionally not compatible with either the wave solder or cleaning operation. If the components are too large, heat-sensitive, or solvent-sensitive, they are assembled at final assembly and manually soldered. A manual cleaning cycle is then necessary to remove flux residues.

Cored solders are also used in the soldering of wires on terminals, connectors, chassis, printed wiring boards, etc. We will cover this process in more depth in Chap. 6; however, special consideration must be given to choice of a flux for this operation. Anytime wires are soldered, caution should be used before employing water-soluble flux. This flux can readily wick up beneath the insulation. Removing the residues is often difficult if not impossible. If residues cannot be removed, use flux that need not be removed, i.e., a mildly activated rosin flux. In some cases cleaning is omitted and provisions are made in designing the process to not clean the assemblies.

BIBLIOGRAPHY

1. Alvin Schneider, *Fluxes Choices for Soldering Printed Wiring Assemblies*, Electronic Packaging and Electronics, July 1979.
2. *Chemical, Fluxes, Solders*, Assembly-Joining Handbook, IPC Joining Materials, June 1978, p 3.
3. Howard Manko, *Soldering Fluxes, Past and Present*, Hollis Engineering Inc., Nashua, NH.
4. Ref. 2, pp 4,5.
5. Phillip Wh. Schuessier, *Water Soluble Flux Technology*, Insulation Circuits, December 1984, p 88.
6. *IPC-S-815*, Rev B. IPC, Evanston, Ill.
7. *The World Book Encyclopedia*, Field Enterprises Inc. Chicago, Illinois Vol. 1, 1948, p. 235.
8. Ref. 7, Vol. 16, pp 8062.
9. Ref. 7, Vol. 10, pp 4335.
10. Ref. 7, Vol. 10, pp 4336.

ASSOCIATED READINGS

1. Dennis Bernier, *Soldering Fluxes*, Kester Solder Co., Chicago.
2. *Examining Solders, Fluxes, and the Soldering Processes*, Insulation Circuits, May 1978.

3. James P. Langan, *How to Select Fluxes for Maximum Effectiveness*, Insulation Circuits, December 1978.

4. Dr. Wallace Rubin, *Non Corrosive vs. Water Soluble Corrosive Flux*, Circuits Manufacturing, March 1980.

5. Laura J. Turbini, Joane G. Eagel, Thomas J. Stark, *IPC-TR-305, A Comparison of Removal of Activated Rosin Flux by Selected Solvents*, Western Electric, IPC Fall Meeting, Sept. 1979.

6. Charles S. McElwes, *Evaluating Organic Fluxes*, Electronic Packaging and Electronics, January 1981.

7. Alvin Schneider, *Water Soluble Organic Fluxes For the Electronic Industry*, Insulation Circuits, February 1979.

8. Dr. Jack Brous, *Water Soluble Flux and Its Effect on PC Board Insulation Resistance*, Electronic Packaging and Electronics, July 1981.

9. Michael Boulos, T. O. Duyck, *Water Soluble Soldering Raises Reliability Rate For Telephone Circuit Packs*, Insulation Circuits, April 1979.

10. Weinberger, D. Audette, *Energies of Activation and Rates of Solder Wetting with Activated Fluxes*, Insulation Circuits, July 1980.

11. Howard W. Markstein, *Synthetic Activated Flux Promises a Combination of Benefits*, Electronic Packaging and Electronics, September 1982.

12. William R. Studnick, Charles C. Foune, *Testing for Corrosivity in Activated Liquid Soldering Fluxes*, Western Electric Engineer, January 1973.

13. William R. Studnick, Charles C. Foune, *Investigation of Selected Water Soluble Soldering Fluxes for Printed Wiring Board Applications*, Western Electric Engineer, October 1975.

14. Howard H. Manko, *Organic Acid Fluxes—A Misnomer*, Circuits Manufacturing, March 1980.

15. Howard H. Manko, *Color, Corrosion and Fluxes*, Electronic Packaging and Electronics, February 1969.

16. *Examining Solders, Fluxes, and The Soldering Processes*, Insulation Circuits, May 1978.

3

Designing the Soldering and Cleaning Process for the Product

3.0 KNOW THE PRODUCT AND ITS USE

To properly design and qualify a process for soldering, one must understand the products or components to be soldered. When designing a product, electrical requirements typically are the major criteria; however, to properly join the parts together by soldering, information on certain mechanical requirements must be identified, specifying the necessary component lead material and surface finishes.

These mechanical requirements differ from product to product; however, with proper definition and qualification, products can be made to solder successfully. Therefore to develop a good soldering process, one must fully understand the components and metals to be joined. In developing this knowledge of component requirements, a matrix can be developed that outlines the best materials and chemicals for a given soldering process. This matrix should include all the mechanical requirements necessary to purchase solderable parts and certify their solderability until the time of soldering or for a given period of storage time.

3.1 DESIGNING THE PRODUCT

Beyond the electrical parameters, many mechanical elements are required to properly design a printed wiring board (pwb) for soldering. Some critical elements are:

- Aspect ratio (board thickness to hold diameter)
- Lead-to-hole size ratio
- Configuration of the solder pads
- Plated surfaces
- Metallic balance over the entire surface of the pwb
- Metallic balance on both sides of the pwb
- Laminate material
- Solder mask
- Component orientation
- Component height off the board

These elements will be discussed in detail later in the text.

In designing a pwb, one must understand the relationship of the solder joint to both the mechanical and electrical properties of the solder termination. The electromechanical function of the solder joint is to hold the component in place during the product's functional life and not deteriorate over time. It must also allow the free flow of electrons without increasing the impedance of the circuit. Mechanically, solder joints fall into two categories:

(1) *Through-hole mounting* implies that the component's leads are mounted into a printed circuit board, through the holes in the board, which can be either plated through or non-plated through. In the case of plated through holes, the electrical/solder connection is made through the entire thickness of the module, whereas in nonplated through holes the solder connection is made only on the side of the module that has the metallic etch lines. Typically that side is referred to as the solder side of the pwb.

(2) *Surface mounting* is as its name implies. The components are placed on the surface of the board and the leads are attached to the metallic traces or etch lines identified as circuit pads, on the same side of the board. In using surface mount components, the components can be placed either on one or both sides of the pwbs. Whichever method is used to create the assembly, certain conditions need to be met for successful solder connections.

3.1.1 Board Fabrication Specification

Typically, plated through holes for both commercial and military applications follow the plating requirements outlined in the specifications stated in references: MIL-P-55640, MIL-STD-275, MIL-P-55110 C, IPC-A-600 [1-4]. The specifications require the following plating processes.

A. Electroless copper plating. Sufficient for electrodeposition.

B. Electrolytic copper plating = 0.001 in. minimum
Tin-lead electroplate = 0.003 in. minimum
Fused electroplate = 0.003 in. minimum

C. Gold plating = 0.000050 in. minimum
= 0.000100 in. maximum

D. Nickel plating = 0.0002 in.. maximum

E. No voids in the plated through-hole barrels

F. Correct pad size and configuration

G. Layer-to-layer registration

H. Minimum thickness of dielectric layers

I. Hole pattern accuracy

J. Annular ring specifications

K. Correct orientation of components on the pwb

Solder mask is another item that must be specified and monitored closely during the process. The original intent of solder mask was to prevent the solder from bridging across etch lines and creating solder shorts. However, with the new technologies available today, solder mask is taking on new roles, such as being used as an insulation resistance protector, impedance control, etc. Therefore, one must understand the solder mask and the materials of the solder mask to properly process it during the manufacturing cycle.

The application and cure of the solder mask is important to the quality of the finished product; therefore, it is critical that it be properly specified and applied to the pwb (Reference IPC-SM-840 for qualification of application [5]).

Special understanding of the chemical characteristics of the base printed wiring board fabrication materials is necessary to determine its compatibility with the assembly process chemicals. Many chemicals react with solder mask in different ways which could be detrimental to the final product. Another issue to be addressed is the curing of the solder mask. Numerous problems occur because the solder mask is not properly cured. Uncured solder mask is not normally found until the product is either wave-soldered or has had surface mount component inserted and reflowed. At that time, it is too late to prevent the problem and corrective action is almost impossible. If the solder mask is inspected for curing at incoming inspection, production problems can be reduced or eliminated.

3.1.2 Component Layout Configuration

The orientation of the components on the board has a direct bearing on the quality of the solder joint. Improper orientation will create solder shorts that are repeatable on each and every pwb and are expensive to remove. This is observed when integrated circuits (ICs), pin grid arrays, and single in-line packages (SIPS) are oriented so that their long axes are parallel to the solder wave. The proper orientation for these types of components is to orient them with their long axes perpendicular to the solder wave. This allows the solder to progressively peel off the component leads as the board is exiting the solder pot, leaving the leads free of any excess solder or solder shorts.

Defects are expensive to repair, and designing products correctly and manufacturing them correctly the first time through the process is the least expensive method of manufacturing. The average standard time to repair a solder short is 15 min; hence if the labor rate is $5.00/hr then the direct cost to remove a short is $1.75. Although this time and cost appear high, it is understandable because it includes the searching, testing, and repairing cost for the defect.

Many solder shorts are caused by improper lead lengths on the solder side of the boards. The maximum length of component leads on the solder side of the boards must be defined in the master drawing of the circuit assembly. This criterion is important, to prevent shorts and electrical failures due to pwbs touching one another or contacting any metal surface once they are in operation.

The layout for surface mount components and the actual configuration of the component leads must be mechanically inspected to certify dimensional accuracy. The pick-and-place process of assembling surface mount components to the pads needs to be accurate. Accuracy is necessary due to the very close spacing of the component leads and the potential mechanical dimensional mismatch between the board and the component. Small leadless surface mount components will realign themselves when the solder reflows due to their buoyancy and surface tension of the liquid solder, whereas the leaded surface mount components will remain where they are placed during the assembly process; therefore, their correct positioning is critical to obtaining a good solder joint. The proper quality and quantity of the solder paste or cream to be deposited must be determined during the process design stages. This is to ensure that the process parameters have been developed and are capable of being followed when the product is introduced into manufacturing.

3.1.3 Component Lead Material

In designing a product, one must review the various lead materials and finishes available that will best suit the needs of the soldering operation. It's important to know that a variety of base lead materials and surface-plating finishes do exist. The soldering of various metal-plating finishes is different from metal to metal, hence incoming solderability testing should be incorporated to determine whether or not these component leads will solder during the assembly-soldering process.

Copper alloy and alloy 42 are the most common lead materials due to their electrical characteristics and their ability to be soldered. Other materials such as nickel alloys and iron are sometimes used due to the mechanical properties required of the component. Since nickel alloys and iron are inherently very difficult to solder, certain solderable finishes must be applied to them to assure good solderability. These solderable coatings are not only applied as protective coatings, but if the base material is non-solderable, a two-step protective coating is usually applied so the new base material will become solderable. Electrodeposits of tin-lead (Pb-Sn), gold (Ag), tin (Sn), cadmium (Cd), and copper (Cu) are some of the typical surfaces applied to preserve solderability. Solder dip and immersion tin can also be applied to protect solderability: however, when these coatings are applied, the base material must initially be solderable [6].

3.2 THE PROCESS OF COMPONENT CERTIFICATION

3.2.1 Component Specification

Once the components and electrical layouts of the printed wiring assembly (pwa) are defined, completing the electrical parameters of the product, specifications must be generated identifying the electrical and mechanical requirements of both the discrete components and pwbs.

Generating specifications involves the following engineering disciplines: component, design, quality, mechanical, electrical, and manufacturing. This engineering consortium, working together in a producibility effort, must define all the properties of the components to certify that they will meet the requirements of the product and the process.

Final component requirements must be identified in a specification to provide the purchasing personnel with the tools to negotiate competitive bids of the necessary parts. These specifications are also used by all incoming inspection departments to certify compliance of the product.

3.3 VENDOR COORDINATION

To qualify materials from any vendors, specific testing must be performed with the materials to ensure their integrity. This testing includes developing the electrical and mechanical tolerances that will be acceptable and conform to the specifications of the final product.

An inspection plan to properly inspect either the outgoing product from the vendor or the incoming product at your given plant must be developed and must include the acceptable quality level (AQL) or parts per million (ppm) definition for incoming product. Acceptable quality level is based on statistical information which will yield a certain quantity of good units based on some sample of inspected units. It defines the risk of accepting some quantity of bad units from a given inspected lot. Presently, however, the need is to produce higher quality products, and therefore, the term AQL is being replaced by the term ppm. Using ppm as the criteria for measurement, the level of quality is defined as a quantity of defective part divided by 1 million. For more information on this subject the reader should review some texts on quality control.

If products are found to be outside the established contractual requirements, they must be rejected and returned to the vendor. If any rework is to be performed to make the parts agree with the specifications, this should be negotiated to certify that reliability of the parts will not be affected.

3.4 QUALITY PRODUCT

Fourteen steps that should be implemented to produce quality product are outlined below. These steps must be taken by top management. Quality of product is the responsibility of everyone, and to produce a quality product the following steps and considerations must be established:

1. Innovation is necessary. Remember that what was good yesterday may not be good tomorrow.
2. Learn a new philosophy. No more mass inspection of incoming product. Buy materials that are right. We cannot tolerate defects coming into the process and product.
3. Discard bad and unusable processes.
4. Reduce the number of vendors.
5. Recognize the true source of trouble.
6. Institute modern methods of training. Define when an operator is properly trained.
7. Improve supervision. Eliminate emotion from processes. Certify that employees know what their jobs are and how to do them.
8. Eliminate materials unsuited for the job. Perform equipment preventive maintenance; don't wail until equipment breaks down before repairing it.
9. Break down barriers between departments. A quality product is produced by a team, not an individual.
10. Eliminate numerical goals. Establish goals of direction, not quantity.
11. Review work standards; estimate assembly cost in advance.
12. Develop massive training programs in rudimentary statistical techniques.

13. Implement training programs in new skills for design changes and new equipment and machinery.
14. Make people secure. Drive out fear from the employees [7].

The implementation of these programs will produce quality product from an understanding of the requirements of the product. These programs were implemented by the Japanese industry in the 1950s from consultations with Dr. Deming, and their success story is still unfolding today. It will take years to fully implement these programs, but procrastination in starting will negate long-range success.

3.5 INCOMING INSPECTION

Incoming inspection was developed to prevent nonconforming material from getting into the manufacturing operation. It is very expensive to plan, develop, implement, operate, and maintain this type of inspection which has at times been called a necessary evil.

New management theories and concepts are driving to push the function of incoming inspection to the manufacturer, with the thought that the responsibility of inspection and quality of product should rest with the manufacturer. Although this is thought to be utopia, it must be performed to properly understand what quality levels are being received for any given product. Continuously monitor quality levels with the vendor while trying not to lose perspective regarding the final results required.

If incoming inspection is performed, the process shown in Table 3.1 is usually employed.

3.5.1 Solderability Inspection

Keeping within the scope of this text, we will confine ourselves to solderability inspection. Solderability, by definition is the ability of the metal surfaces to accept solder without displaying characteristic signs of nonwet and/or dewet conditions. We will

Table 3.1 Incoming Inspection Process

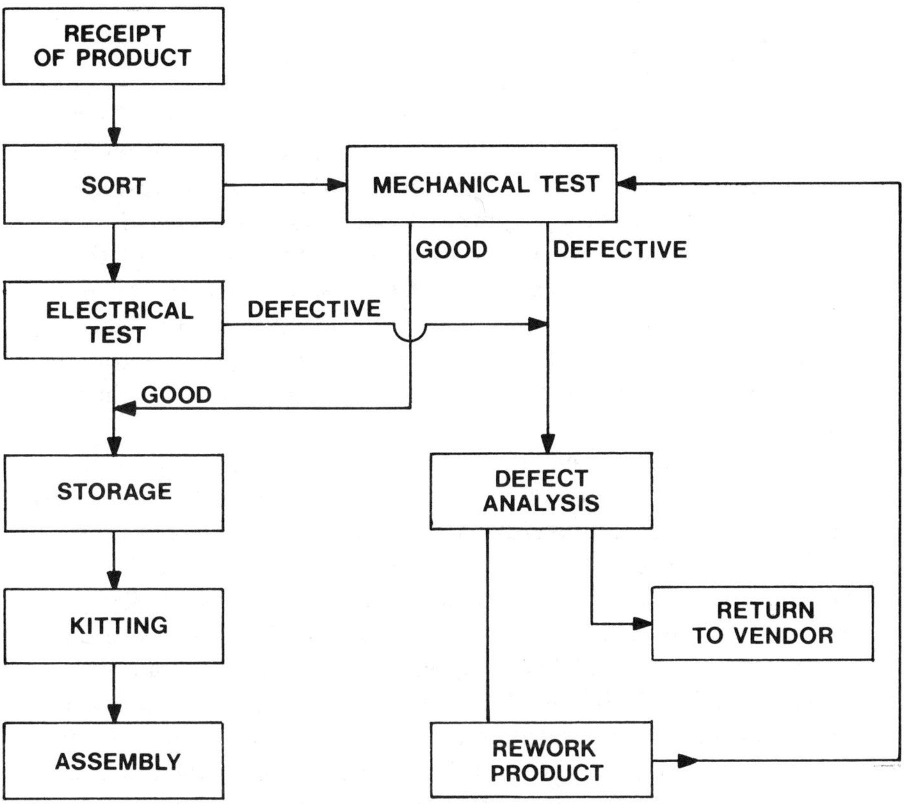

develop the technical requirements in Chap. 3; however, for the purpose of this chapter, let us review the methods, tools, and specifications for performing an inspection.

In checking for solderability of components and boards, the objective is to measure or observe the qualitative or quantitative results of the following elements:

- o Wetting forces
- o Wetting time
- o Meniscus rise

The tests used to accomplish these measurements are identified in Table 3.2. These tests define whether or not surface finishes are solderable and whether or not the base materials of the test coupons have been properly prepared. Coupons are samples of printed wiring boards which are used in the performance of the tests. It must be noted that some of these solderability tests may be destructive to the printed wiring boards, rendering the test module useless for manufacturing. Hence, to compensate for this loss of product, extra materials will be needed at incoming inspection.

A method to prevent this extra cost of destroying product is to procure test coupons on which the test can be performed. Test coupons are fabricated as part of the printed wiring board fabrication process and are, therefore, theoretically identical. Solderability tests performed on the test coupons should and will represent what will actually occur to the printed wiring board during the process. The use of test coupons is also an aid in maintaining records of solderability or other mechanical requirements of the printed wiring board. Plating measurements, epoxy smear, intermetallic thickness, alloying content, and cleanliness levels are all items that can be referenced from the basic test coupon.

3.5.2 Test Procedures

Tests must be performed on both components and pwbs, as prescribed by the manufacturing specifications, and some tests will be qualitative and others will be quantitative in measuring the results. Through the years, visual examination was typically the only method to certify solderability, and the resultant joint or surface had to be smooth, shiny, and show no signs of nonwet or dewet. Wetting or "to wet" is defined as the time it takes for the coupon to reach soldering temperatures and allow the solder to

Table 3.2 Test Procedures

Test	Standard
Edge dip	○ IPS-S-801 Edge Dip Solderability Test for Printed Wiring Boards ○ IPC-S-804 Solderability Test Method for Printed Wiring Boards ○ IPC-S-805 Solderability Test for Component Leads and Terminations ○ Mil-Std-202, Test Methods for Electronic and Electrical Parts ○ RS-319-A Solderability of Printed Wiring Boards ○ EIA-RS 178 Solderability Test Standards ○ EIA-RS 319 Solderability of Printed Circuit Boards
Rotary dip	○ IPC-S-804 Solderability Test Methods for Printed Wiring Boards ○ IEC 68-2-20, Part 2T Environmental Test Procedures (Soldering) See Fig. 3.1
Globule	○ IEC-68-2-20, Part 2T, "Environmental Test Procedures" (Soldering) ○ IPC-S-805 Solderability Test for Component Leads and Terminations
Surface Tension balance test	○ Mil-Std-883, Method 2002, Solderability ○ IPC-D-805 Solderability Test for Components Leads and Terminations

flow in a continous film over its entire surface. If the solder does not create a continuous film over the surface, the resultant condition is known as dewetting. If the solder completely pulls away from the base material, that condition is known as nonwetting of the base material. However, at times these results cannot be

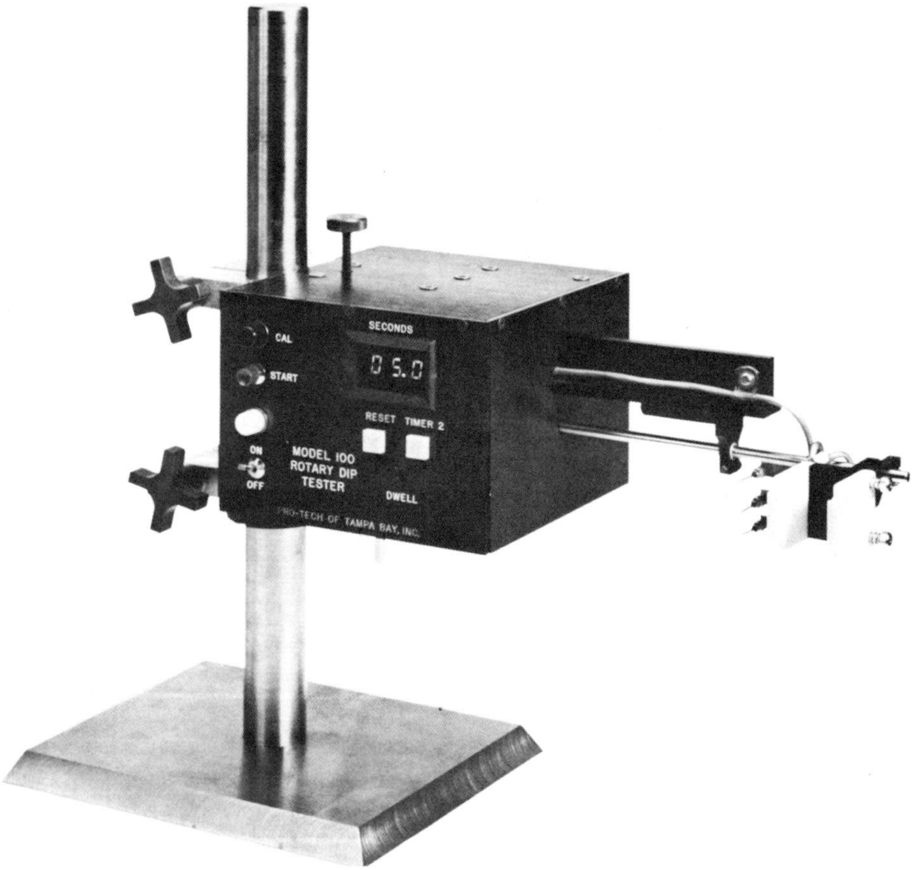

Figure 3.1 Rotary dip tester. The purpose of this tester is to replicate the wave-soldering process for performing soldering tests on printed wiring board coupons. The rotary motion simulates the product entering and exiting the wave solder pot at the correct angle. The dwell time in the solder is also controlled to meet the requirements established by the process and product designs. (Courtesy of Pro-Tech of Tampa Bay Inc.)

exactly correlated with manufacturing processes. So quantitative tests were developed that can be used for correlation purposes.

Quantitative edge dip, solderability tests were developed to identify the time it took a coupon to wet and measure the rise of the meniscus along the vertical edges of the test coupon. Measurement of the solder rise identifies the time it takes for the solder to wick or rise up the component lead or plated through hole. This phenomenon occurs because the wetting forces applied by the molten solder are in the downward direction. The rise of the meniscus is therefore identified as this wicking action on the part being soldered.

Under certain conditions test coupons are chemically prepared or aged prior to solderability testing. The samples are prepared for the solderability test by a cleaning process, then a surface preparation with aqueous benztriozole solution (BTA).

The cleaning process includes a solvent cleaning in either chlorinated or fluorinated solvent. These solvents will remove nonpolar contaminants such as finger prints, greases, and oils. Then, to remove the metallic oxides, an etch cycle in 25% by weight aqueous ammonium persulfate, then a rinse in 0.2% by volume solution of sulfuric acid, with a final dip in acetone with air dry is utilized. The BTA treatment is then applied to create a surface condition on the surface of each coupon. This cleaning cycle is applied to eliminate any variables on the test coupon [9].

Steam aging is another process used to simualte storage of the devices in a typical manufacturing environment. Although many methods exist to artificially age the coupons, one method that is coming to the forefront is the steam-aging test. This test consists of suspending the test samples over boiling distilled water for a period of 24 hr. It is being used more often because of the ability to age the samples within a reasonable time limit and correlate this to the actual storage environment, predicting accurately the solderability of the parts.

3.5.3 Test Method Definition

Solder Spread Test

This is a multifunctional test used to measure the effectiveness of the flux in preparing the surface to be soldered or measure the ability of the material to be prepared by the flux. Either the flux type or the coupon surface can be held constant while the other is the variable during the tests. This will define the cleaning ability of the flux in enhancing the ability of the solder to wet the surface.

The test requires the use of a localized heat source, like a hot plate, heated to predetermined temperature, fluxing of the coupon, then the placement of the solder onto the test sample. The results are found by either measuring the total area of spread of the solder or measuring the thickness of the solder on the test coupons. This permits a comparison to be made between samples, allowing some quantitative data to be gathered.

This test can also be used to measure the effectiveness of fluxes to promote wetting and wicking, and to determine what solder reflow temperatures are best for various fluxes and solder alloys.

The Edge Dip Test

This test is at times called a vertical edge dip test and consists of dipping a coupon in flux, immersing it in molten solder at a prescribed temperature and entry rate, allowing it to dwell in the solder pots, and removing it at a given exit rate. The sample is then observed for flow of solder and any characteristic wetting defects. This test is used for component leads and laminate materials.

The Globule Test

Primarily used to measure the solderability of component leads (see Fig. 3.2), this test consists of placing a given size solder pellet and melting it on a heated post (anvil). Once melted, the solder globule is bisected with the lead to be tested. As the lead is heated, the solder pellet wraps itself around the lead because of

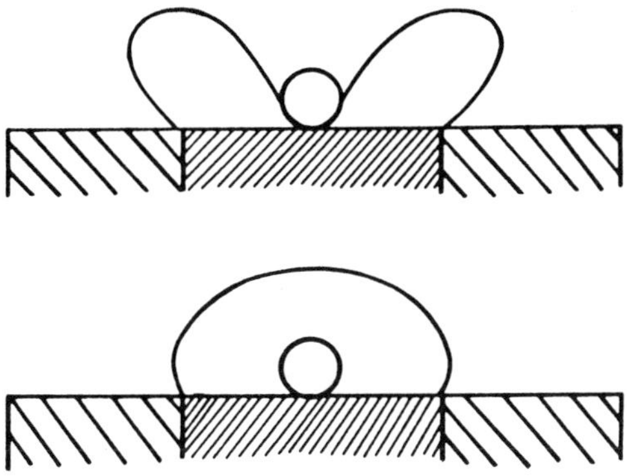

Figure 3.2 The multicore solder globule test. (Courtesy of Multicore Solders Ltd.)

the wetting action ofthe solder on the component lead. The period of time to wrap itself around the lead is measured and recorded identifying the solderability time of that particular component. With modifications, this type of equipment can also be used to measure the time rise of solder in a plated through hole.

The size of the anvil, the heated post on which the solder pellet is placed, is determined by the diameter of the lead or the plated through hole to be tested, the implication being the amount of thermal energy which will be required to create the solder joint. The larger the lead or the plated through hole, the more metallic mass is involved, thereby requiring more thermal energy to heat the coupons or samples to soldering temperatures. The time to wet can be measured electrically with readily available standard equipment. It can also be observed visually and measured by a stop watch. Although the stop watch is not as accurate as the electronic method, proper training will yield a consistency of readings that can develop measurement ratios. The measurement ratio is the correlation between the actual process soldering time and the time to solder using the solderability tester.

The measurement ratios can be analyzed statistically to certify the solderability of the components. Such information is used to develop vendor/user communications that are important in the manufacturing of quality products.

Meniscus Rise Rate

Meniscus rise rate was developed by Philips, Eindhoven, and the General Electric Co. (GEC) of Great Britain (see Fig. 3.3). The test is also known as surface tension balance and provides the ability to quantify both wetting and meniscus rise time. Typically, a test coupon is attached to a transducer spring balance. The test cycle consists of bringing a solder bath into electrical contact with the fluxed test coupon. Once electrical contact is made between the solder and test coupon, the solder pot rises to a predetermined depth. Dwell time in the solder pot is fixed by particular test parameters, and the entry and exit rate are fixed by the system itself.

> Initially, a specimen vertically immersed tends to carry the surface downward with it and surface tension resists the immersion with a measurable force. As the sample becomes wetted, the edge of the meniscus formed gradually reverses direction and ultimately tends to rise on the surface of the sample. When this happens, the forces acting on the sample also reverse and tend to pull the sample downward. The test instrumentation measures and records these changing forces as a function of time [10].

The meniscus rise test is used for laminate material and component leads.

In all these tests, specific variables can be altered to inspect for different elements of printed wiring boards and components. Therefore, different fluxes, solder alloys, and soldering times can be utilized with these tests to properly define the quality of the incoming product. This is accomplished by creating only one variable during the tests and maintaining all other test parameters constant. The variations in the results can then be attributed to the variables of the experiment.

Other processes should also be used at incoming inspection to identify the quality of incoming solder and chemicals. Solder

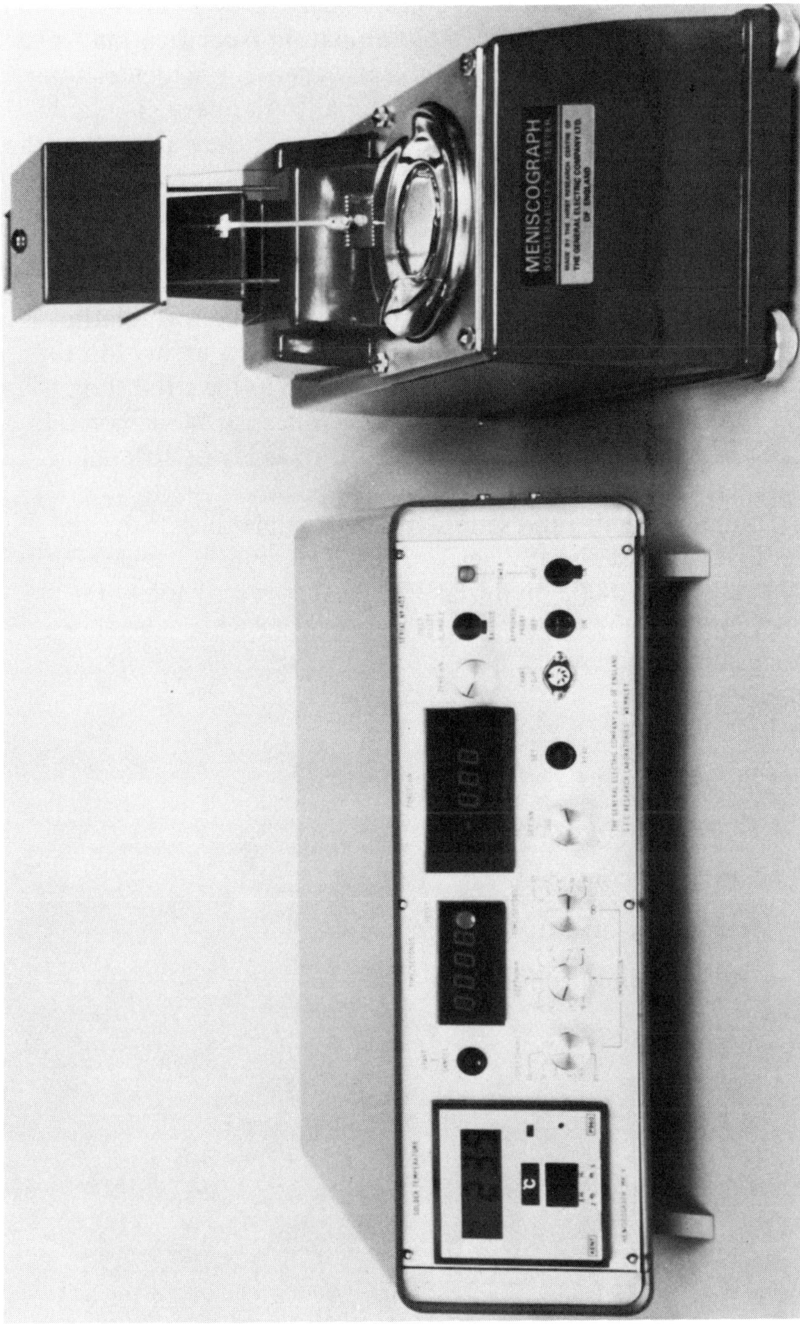

Figure 3.3 Meniscograph solderability tester. (Courtesy of Hollis Automation Inc.)

should always be received with an assay certificate of analysis identifying its elemental metal contamination (specification QQ-S-571 and ASTM-B-32, 76). See Tables 3.3 and 3.4, which can be used for correlating the incoming material to industry standards.

Fluxes can also be checked to certify that they are within the established guidelines of the vendor. Specific gravity measurements and color differentiation are two of the most popular methods for inspecting fluxes. Typically all fluxes are proprietary blends of various chemicals, with the vendor only identifying the major ingredients and never the percentages or the formulation process. Therefore, inspection of the actual elements within the flux are limited and any special efforts made to have the flux analyzed through a laboratory would not necessarily be worth the cost of the work. Hence lab analysis of fluxes is not normally performed with incoming flux products.

To determine if fluxes are within the guidelines of the process, other tests are usually performed to measure the corrosivity and effectiveness of the material. Rosin fluxes are usually inspected per MIL-F-14256 [11]. These tests to which the fluxes must be submitted are:

Materials

- Solids content
- Color and transparency
- Flow and consistency
- Resistivity of water extract
- Chloride and bromide test

Fluxing action

- Solder spread test
- Solder pool

Copper mirror test

- Dryness
- pH

Table 3.3 Federal Specification (Adapted from QQ-5-571).

Composition	Tin (%)	Lead (%)	Antimony (%)	Bismuth max. (%)	Silver (%)	Copper max. (%)	Iron max. (%)	Zinc max. (%)	Aluminum max. (%)	Arsenic max. (%)	Cadmium max. (%)	Total of all others (%)	Approximate melting range (°C) Solid	Liquid
Sn 96	Remainder	0.010, Max.	–	–	3.6 to 4.4	0.020	–	0.005	–	0.05	0.005	–	221	221
Sn 70	69.5 to 71.5	Remainder	0.20 to 0.05	0.25	–	0.08	0.02	0.005	0.005	0.03	–	0.08	183	193
Sn 63	62.5 to 63.5	Remainder	0.20 to 0.05	0.25	–	0.08	0.02	0.005	0.005	0.03	–	0.08	183	183
Sn 62	61.5 to 62.5	Remainder	0.02 to 0.05	0.25	1.75 to 2.25	0.08	0.02	0.005	0.005	0.03	–	0.08	179	179
Sn 60	59.5 to 61.5	Remainder	0.20 to 0.05	0.25	–	0.08	0.02	0.005	0.005	0.03	–	0.08	183	191
Sn 50	49.5 to 51.5	Remainder	0.020 to 0.050	0.25	–	0.08	0.02	0.005	0.005	0.025	–	0.08	183	216

Table 3.4 Federal Specification (%) (ASTM B 32-76).

Alloy grade	Tin desired	Lead nominal	Antimony min.	Antimony desired	Antimony max.	Silver min.	Silver desired	Silver max.	Bismuth max.	Copper, max.	Iron max.	Aluminum max.	Zinc max.	Arsenic max.
70A	70	30	–	–	0.12	–	–	–	0.25	0.08	0.02	0.005	0.005	0.03
70B	70	30	0.200	–	0.50	–	–	–	0.25	0.08	0.02	0.005	0.005	0.03
63A	63	37	–	–	0.12	–	–	–	0.25	0.08	0.02	0.005	0.005	0.03
63B	63	37	0.20	–	0.50	–	–	–	0.25	0.08	0.02	0.005	0.005	0.03
60A	60	37	–	–	0.12	–	–	–	0.25	0.08	0.02	0.005	0.005	0.03
60B	60	40	0.20	–	0.50	–	–	–	0.25	0.08	0.02	0.005	0.005	0.03
50A	50	50	–	–	0.12	-	–	–	0.25	0.08	0.02	0.005	0.005	0.03
50B	50	50	0.20	–	0.50	–	–	–	0.25	0.08	0.02	0.005	0.005	0.03
2.5S	0[a]	97.5	–	–	0.40	2.3	2.5	2.7	0.25	0.08	0.02	0.005	0.005	0.02
1.5S	1[b]	97.5	–	–	0.40	1.3	1.5	1.7	0.25	0.08	0.02	0.005	0.005	0.02
95IA	95	0.20 max.	4.5	0	5.5	–	–	–	0.15	0.08	0.04	0.005	0.005	0.05
96IS	96	0.20	0.20	–	0.50	3.6	4.0	4.4	0.15	0.08	0.02	0.005	0.005	0.05

[a]Tin maximum, 0.25%.
[b]Permissible tin range, 0.75-1.25%.

Synthetic fluxes are becoming more common throughout the commercial electronics industry. New tests are being developed to certify their effectiveness, consistency, and reliability under a variety of production processes and working environments.

Special tests that are being developed in addition to those above for certification of these materials are:

- Insulation resistance
- Environmental corrosion
- Thermal cycling under operating bias voltages
- Cleanliness, or ability to be removed, using various cleaning solutions
- Laminate degradation
- Component compatibility

3.6 STORAGE

Proper storage, packaging, and handling are all very important to prevent damage and preserve the solderability of the components and modules. Solderability of all surfaces deteriorates over time, and this phenomenon is accelerated by long storage, contaminated and corrosive atmospheres, and improper packaging materials. Material storage systems have to be developed to incorporate first-in/first-out (FIFO) concepts. Many have been designed improperly creating a last-in/first-out (LIFO) which creates a first-in/stays-here (FISH) system, which allows unacceptable aging on the shelves. Proper planning of a storage area incorporting FIFO systems and easy access to all products is important to the success of the manufacturing process quality.

3.6.1 Contaminated or Corrosive Environments

In laying out the flow of product through many manufacturing plants, storage rooms have had a tendency to be included as a last thought in the design process. Storage areas or warehouses must be designed with the proper environmental controls to prevent the deterioration of the product.

Corrosive environments exist where gas-powered fork lift trucks are used. Also, any storage that is close to the shipping and receiving docks where fumes of the ongoing traffic are allowed to enter will cause corrosion of the product. An environment of clean air-controlled temperature and humidity is a prerequisite for storage of any components that will be soldered.

3.6.2 Handling

All personnel who handle components to be soldered should be instructed on how to avoid contaminating solderable surfaces. The personnel must also be made aware of potential lead ingestion that can occur when handling solder-coated parts. An example can be seen by wiping a solder-coated component lead with a paper towel and observing a black line on the towel. This line is lead oxide which, if left on unwashed hands, could be ingested when eating. Another issue of concern is packaging material used with static-sensitive devices. Improper packaging material can build up static charges that are detrimental to C-MOS, MOS, ROMS, PROMS, and other memory devices. Static protection must be planned in the initial stages of establishing a manufacturing line.

Vendors must also be made aware of potential static problems with certain components. All static-sensitive components must be handled in containers that are static-free.

Material handling during the process is another step that requires planning prior to implementation. An overused phrase is apt: "One must define what one wants prior to starting a project." This definitely applies to in-process handling containers. Concerns regarding lot size, board size, weight of handling container both empty and full, static protection, automatic loading and unloading vs. manual loading and unloading are all issues that must be addressed prior to designing the proper containers. Also, the space required to store empty containers must be considered; varying work loads could have significant impact on container design criteria.

3.6.3 Packaging Materials

If possible, all parts should be packaged so there is no contact between any two components. This prevents solder from smearing, damage to component leads, and contamination from spreading between parts in the bins and on the shelves.

Sulfur-Free Packaging

All packaging must be sulfur-free. Cardboard boxes, paper bags, and elastic bands if not specified to be sulfur-free will contain large amounts of sulfur. Sulfur will codeposit on the surfaces of all the materials within the confines of the containers. These sulfur coatings are difficult to remove with standard fluxes and should be prevented.

Adjustable Containers

From past experience, handling containers should have capabilities that allow for adjustment to various board sizes. And, because damage to pwbs can occur if one container rests on the boards in the container beneath it, each container should be designed to fully protect its contents.

Component Packaging

Components are handled somewhat differently due to their size and quantity. Loose components, as well as reeled or taped components, are available from all vendors. However, the assembly procedures utilized will define how the components should be purchased and packaged.

Bulk Packaging

The packaging of loose components is probably the main cause of components damage. Some components require special handling due to polarity and this augments the loose packaging problems encountered with these types of components. Another style of component that requires special care is surface-mounted components. These components are typically in a leadless configuration for discrete components and either "J" bend or gull wing flat pack configuration for multileaded active devices.

Packaging and handling of such components will require more planning due to special characteristics resulting from component size. The passive devices can be ordered either loose or in 8- and 16-mm tape. The tape configuration can also handle SOT (small-outline transistors) and SOIC (small-outline integrated circuits). These tapes offer a benefit not normally found in packaging designs: the protective tape lid that is applied to the tape. This now allows complete protection of the device in its unique container within the tape, and it also prevents the components from touching and damaging one another.

Therefore, in planning the packaging of components, consider the process and the process steps necessary to prepare the components for assembly. If assembly of the components is to be automatic, then taped component reels are best; if manual assembly is the method used, then single components may be sufficient. In any case, consider the following:

- Taped
- Loose
- Preforming
- Assembly
    - Automatic
    - Manual
- Packaging material
- Static sensitivity

3.7 BOARD ASSEMBLY

Regarding through-hole components, in considering the assembly function, the process is based on placing components into the holes of the pwbs or on the surface pads for surface-mounted types of components. The major criteria that must be considered for plated through-hole boards are the lead to hole size and aspect ratios. These ratios have a multiplicity of special factors such as: drill size, plating thickness, surface finish, use for automatic inserted component, use for manual assembly of components. Typically, the ratio is designed for ease of automatic insertion; however, a large lead to hole size ratio may result in the solder falling out of the hole during soldering.

3.7.1 Aspect Ratio for Through-Hole Components

Aspect ratios relate to the hole size vs. board thickness. It is usually around 1:2, e.g., 0.0350 in. hole to an 0.062-in.-thick board. Lead to hole size ratio is quite different. The hole should be no larger than 0.020 in. greater than the lead diameter or cross-section diagonal for rectangular leads. The minimum size of the plated through hole or unplated through hole in a printed circuit assembly should be 0.008 in. larger than maximum lead diameter. Therefore,

Maximum clearance for component leads = 0.020 in.

Minimum clearance for component leads = 0.008 in.

Again, it should be noted that the larger the lead to hole size ratio, the larger the probability of solder drop out from the plated through hole, and the smaller the clearance, the more difficult it will be for the solder to wick up through the hole and create a good solder joint.

3.7.2 Surface Mount Components

The assembly of surface mount components requires different techniques in component placement and soldering. There are four major sytyles of surface mount components:

1. Chip components, i.e., capacitors, resistors, and diodes

2. SOIC (small-outline integrated circuits)
   SOT (small-outline transistors)
3. Leadless ceramic devices
4. Leaded devices

Exact positioning of these devices is critical due to the unique lead configurations and pad layouts being generated for printed wiring boards. The lead configurations have either an 0.050- or 0.025-in. center-to-center distance, which necessitates critical positioning of the devices onto the pads of the board surface. Additionally, with surface mount components, there is no mechanical security of the device to the board such as through-hole components process. Hence, the only security that the components have during the assembly process is the tackiness of the solder paste onto which the devices are placed prior to reflow, unless adhesives are used to secure the components onto the pwbs before soldering.

The process of assembling surface mount components is a critical consideration in the design of electronic assemblies. Since massive reductions in board surface area are being realized on printed wiring boards, more surface mount components will be used to increase the functionality of the board or system. The key word is miniaturization and since the devices are getting smaller, they are becoming more difficult to place and solder.

We will delve into the process later in the text; however, it must be noted that since surface mount components are not soldered with the same techniques as the through-hole components that have been used in the past, certain new functions are being developed to place the solder cream or solder paste in the exact position, and then place the components exactly into location.

Automatic solder paste-dispensing systems and robotic arms are being utilized to assemble surface mount components onto printed wiring assemblies. The placement and security of the devices is of prime concern during the assembly process. Many handling operations and movements through the facility may disturb the location of the devices in relationship to their pads on the board and perhaps cause defects that will be difficult to repair.

The solderability of these components is more critical than the through-hole-type components due to the miniaturization of the solder joint itself. High-density through-hole mounting has approximately 35–40 joints per in.^{2} of module area. Using completely surface-mounted component populations, this density will rise to 80-90 joints per in.^{2}. Therefore, it will be mandatory that proper design of both the components and the process be developed to eliminate the inspection of each and every solder joint. If this does not happen the inspection cost will potentially quadruple and eliminate some of the cost advantages of using surface mount components.

3.7.3 Clinch Leads vs. Straight-Through Lead Attachment

In clinch leads, the lead is bent over on the solder side of the module. This is performed for a variety of reasons, the largest of which is to prevent the components from falling out prior to the soldering process. Military and commercial specifications identify the requirements necessary to create a sound solder joint [1-5].

In designing a printed wiring board and an assembly process, the need to clinch must be made known early in the design planning stages. If not properly identified, improper machines and processes could be put in place that are not conducive to good assembly practices.

All of this falls into the category of solder joint design for strength and electrical adequacy. In the strength categories, the tensile strength of solder is more critical than the tensile strength of the leads. Solder joint failure mechanisms can be fractured due to constant load and thermal cycling of the devices. Another factor is the reliability of the plated through hole and its ability to adhere to the laminated material.

3.7.4 Cutting of Component Leads

Knowing the product and understanding the requirements are major criteria for fulfilling this task. A major question that must be answered prior to establishing this process is, is exposed metal on the tips of component leads acceptable after the soldering operation? Military applications through the years have never allowed exposed metal on the tips of the component leads. The method utilized to prevent this is to cut the leads prior to the soldering operation or utilizing a solder-cut-solder operation, where the leads are cut after the first soldering operation and resoldered to cover the tips of the leads and correct any fractured joints that may have been created during the lead-cutting operation. The majority of commercial electronics allow exposed base metals on the component tips after soldering.

3.7.5 Mass Lead Cutting

Whether performed prior to or after the wave-soldering operation, mass lead cutting is a critical step in the manufacturing process. The leads must be secured prior to mass lead cutting. This can be handled with a wax wave system whereby wax is used to hold the components in place. This wax can but does not necessarily have to have flux as a major ingredient. If the wax has no flux, then a fluxer is necessary prior to the wax application.

With this system, a cooling station is placed prior to the auto cutting wheel of the mass lead cutter. This wax is then transformed into a solid mass that will support the component lead during the mass lead-cutting cycle. Another method is the solder cut solder process, where the solder is used as an anchorage for the component leads. Although the solder is stronger, the potential for cracking the solder joint still exists and must be reviewed constantly to certify wheel sharpness and replacement characteristics.

3.8 SOLDERING TECHNIQUES

Soldering is the process of joining two metals using a third metal. The third metal used to perform this task in the electronics field is normally a tin-lead alloy called solder.

In designing a process for a product, as was mentioned at the beginning of this chapter, it is important to know the limitations of the soldering operation be it either a manual or an automatic process. This section will discuss briefly the soldering processes that can be utilized for creating the solder joints between the components and the printed wiring board.

3.8.1 Manual Soldering

The tools needed for manual soldering are as follows:

1. Soldering iron, of predetermined temperature
2. Cored wire solder
3. Wet sponge
4. Soldering aid
5. Holding fixture for the board
6. Solvent for cleaning the flux

The assembly area that will be used for the manual soldering operation must be designed with the operator in mind. The work area should be designed with the following concepts:

1. Well lit, with no glare spots
2. Proper height of chair and bench, adjustable for various operators
3. Proper place for all the tools and materials
4. Free of nonessential materials
5. An appealing color to prevent eye strain

6. Well grounded to prevent ESD (electrostatic discharge) problems with electrostatic-sensitive components
7. Vented to remove noxious fumes
8. Storage area for operator hand tools and personal belongings

The safety of the operators must also be considered when designing a work bench. The following items must also be taken into considerations when designing the process:

1. No foods or drinks should be allowed.
2. Safety glasses must be worn at all times.
3. Tools must be in good working order.
4. Broken tools must be replaced.
5. Aisles should be kept clear of all boxes.

Manual soldering, as its name implies, is performed with a hand tool or soldering iron that is brought to the connection and utilized to transfer the heat into the metals to be joined. Once the metals are heated to soldering temperatures (361°F for eutectic solder), the solder is then applied to the joint. As the heat from the solder iron transfers to the solder being applied, the internal core of flux in the solder starts to flow onto the metals to be joined and prepares them for solder operation. The function of the flux is to remove the metallic oxides on the surfaces of the components to be joined and create a protective film over all the metallic surfaces to prevent the reoxidation of the surfaces. The solder, once it reaches soldering temperatures, flows onto the components that have been fluxed and joins those metals into a continuous solder joint.

We will discuss the details of the manual soldering operation. This will include the process of choosing the right soldering iron for the job and the correct solder. It will also describe how to perform the operation correctly to prevent damage to the components and the printed wiring board.

3.8.2 Automatic Soldering

Automatic soldering or mass soldering techniques are used extensively throughout the world for the production of printed wiring assemblies also called modules. This process was created many years ago to improve the quality and consistency of the solder joint. Mass soldering consistency is predictable if the process is designed correctly.

Until recently the term automatic soldering implied two basic methods of soldering. One was the wave-soldering system, the other was drag or static pot soldering. However, with the introduction of higher density packaging and surface mount components, new automatic processes are performing the soldering operation. Among such systems are laser soldering, vapor phase or condensation soldering, and infrared soldering to name a few.

In designing the mass soldering process, one must understand the product, primarily because the soldering process is the harshest process to which the printed wiring board will be exposed during the manufacturing cycle. It is important to understand the facets of the process such as metallurgic functions of the solder and the chemistry of the fluxes and tinning oils in the process, and their interaction with the printed wiring board at the soldering temperatures.

Knowledge of the metallurgic and chemical functions as well as of raw board fabrication process will increase the value of any engineer responsible for the process.

The ability of components to withstand high temperatures and cleaning solvents are two of the major areas that must be investigated to understand the total soldering process. We will deal with all of this in detail in a chapters 4, 5, and 7; however, it is important to understand the process while designing the product.

Process specific issues that must be addressed during the design cycle are the size of the equipment itself and the cleaning solvent to be used to remove the contaminants from the finished assembly. Board sizes play a major part in defining how large or small a soldering system should be. Also the quantity of product that will be produced with the equipment is important, primarily because of maintenance and care of the system itself. Soldering equipment requires maintenance on an ongoing basis, and to produce quality product, it is important that it be performed on a regular basis. Waiting until the system breaks down before doing repairs will cause delays in shipment and increase product cost.

Therefore when designing a product it is important to define what kinds of tools manufacturing has to use to produce the product. If the process is not available, then information must be developed to create a process for the fabrication of the product.

3.9 SPECIFYING CAPITAL EQUIPMENT

In specifying capital equipment to perform the soldering operation it is important to define the product to be soldered, the parameters required for the functionality of the product, and the quality required by customers. This information must be provided prior to the qualification of the process.

The process must be qualified by the customer to certify that the equipment is producing product of an acceptable quality. This may sound confusing when the assembly line is developed; however, if done properly at the beginning, then the process will have been developed with the product in consideration.

3.10 PROCESS CONSIDERATIONS FOR SURFACE MOUNT COMPONENTS

If surface mount components are used in the product, then many other issues must be resolved prior to going into production with the product. Surface mount components require the solder to be in place prior to the component placement operation. This is much different from the mass soldering process, where a minimum of 100-1000 lb of solder is applied over the entire bottom surface of the board and expected to adhere only where there is a metallic connection. In the case of surface mount components, solder is applied prior to the assembly of the components; and the components are then placed into the solder in either a cream or paste form. Once the surface mount components are in position, the assembly is submitted to a high-heat environment to reflow the solder cream or paste material.

Although there are many issues of concern during the assembly process, the issues discussed in this chapter must be reviewed in detail to certify that all the process steps have been reviewed and are in control. A controlled process will produce quality product consistently; an uncontrolled process may produce quality product sporadically. Therefore, it is important to design the process with the controls necessary to produce the quality product that customers will buy and use.

BIBLIOGRAPHY

1. MIL-P-55640, Printed Wiring Boards, Multilayer (Plated Through Hole).

2. MIL-STD-275, Printed Wiring for Electronic Assembly.

3. MIL-P-55110 C, Printed Wiring Boards, 26 April 78, Amendment 4, 29 June 82.

4. IPC-A-600, Acceptability of Printed Boards.

5. IPC-SM-840, Qualification and Performance of Permanent Polymer Coating (Solder Mask) for Printed Boards.

6. *Quality Assessment, Component Lead Solderability*, Circuit World, Vol. VI, (No. 1, 1979), p 42.

7. Dr. W. Edward Demings, *Conference of Productivity and Quality Control*, IPC Technical Seminar (April 1981), Washington, D.C.

8. *Solderability Testing*, Tin Research Institute Publication M-211 (August 1982).

9. C. A. Mackay, P. E. Davis, *A Compilation of Solderability Values as Measured by Different Solderability Test Methods*, Tin Research Institute.

10. Dallas Erickson, *Solderability Test Systems*, Electronic Packaging and Production (May 1980), p 98.

11. MIL-F-14256, Flux, Soldering, Liquid.

ASSOCIATED READINGS

1. Armin Rahn, *Protecting the Solderability of DIP Leads*, Circuits Manufacturing, March 1983, pp 68-71.

2. G. Clarmo, *Storage of P.C. Boards*, Swedish Institute of Production Research Eng., date unknown.

3. George Messner, *Packaging High Speed Circuits*, Circuits Manufacturing, February 1983, pp 56–61.

4. James Wheeler, *Eliminating Printed Circuit Design Problems*, Insulating Circuits, May 1979, pp 36-38.

5. Howard W. Markstein, *Consideration in Producing High Density P.C. Boards*, Electronic Packaging and Production, March 1981, pp 46-60.

6. Charles J. Bartlett, Leonard N. Schoenberg, *PCB Lines and Spaces: How Fine*? Circuits Manufacturing, April 1981, pp 54-59.

7. Charles W. Jennings, *Minimum Spacing for PCB Conductors*, Circuits Manufacturing, February 1979, pp 18-24.

8. Howard H. Manko, *CAM for Wave Soldering*, Circuits Manufacturing, January 1983, pp 32-33.

9. H. Terrance O'Rourke, *Controlling Wave Soldering Through the Use of Time/Temperature Instrumentation*, Insulation/Circuits, October 1978, pp 37-29.

4

Automatic Soldering Systems

4.0 INTRODUCTION

The purpose of this chapter is to define the various sections of mass soldering systems and describe their functions and operations in an individual fashion. See Figs. 4.1-4.3 for total wave-soldering systems.

Automating the soldering process was started in order to reduce the cost of creating solder joints. Although many problems were encountered and solved by automating the system, one of the major advantages was the consistency developed among all the solder joints created. By using this process consistently, the quality and reliability of the product was improved many times.

4.1 SECTIONS OF THE AUTOMATIC SOLDER SYSTEM

Soldering is based on a metallurigic function and chemical preparation of the surfaces, and it is usually incorporated as part of a total soldering system. Therefore, for the soldering system to perform exceptionally well, all the variables must be considered and understood prior to and during the soldering process.

All automatic soldering systems have the following basic sections:

- Fluxer
- Preheater
- Solder pot
- Conveyor

Various solder wave configurations are offered as options by each manufacturer, but the overall soldering concept is similar. Other elements offered as standard or optional equipment are:

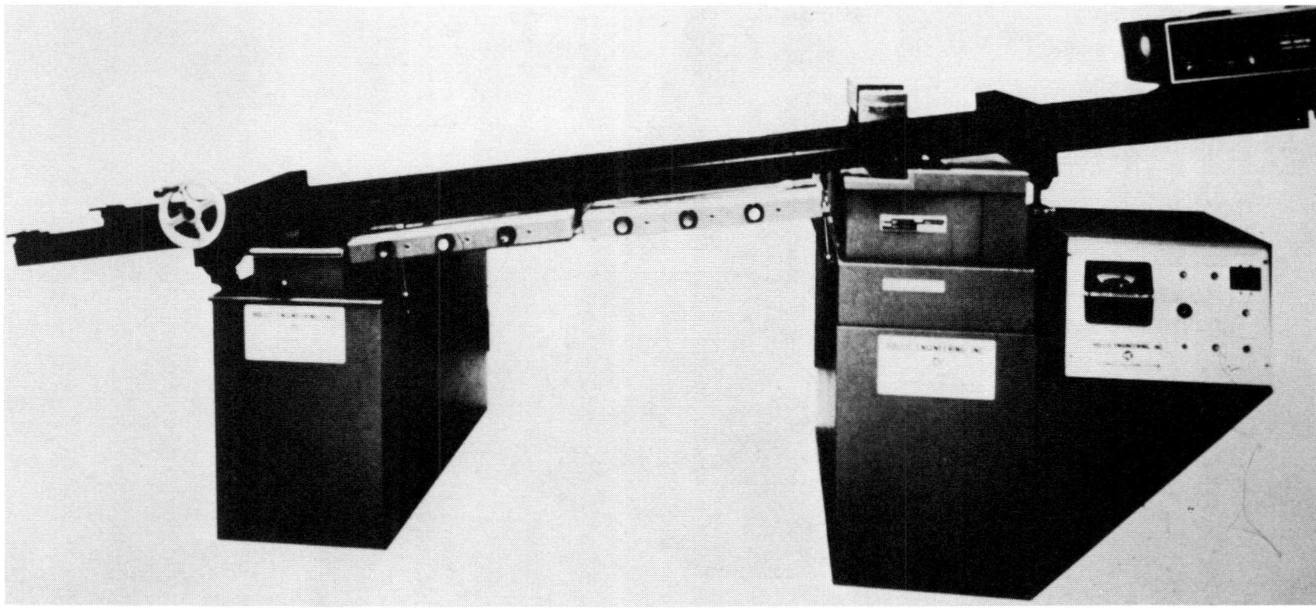

Figure 4.1 Hollis TDC wave solder system. (Courtesy of Hollis Automation Inc.)

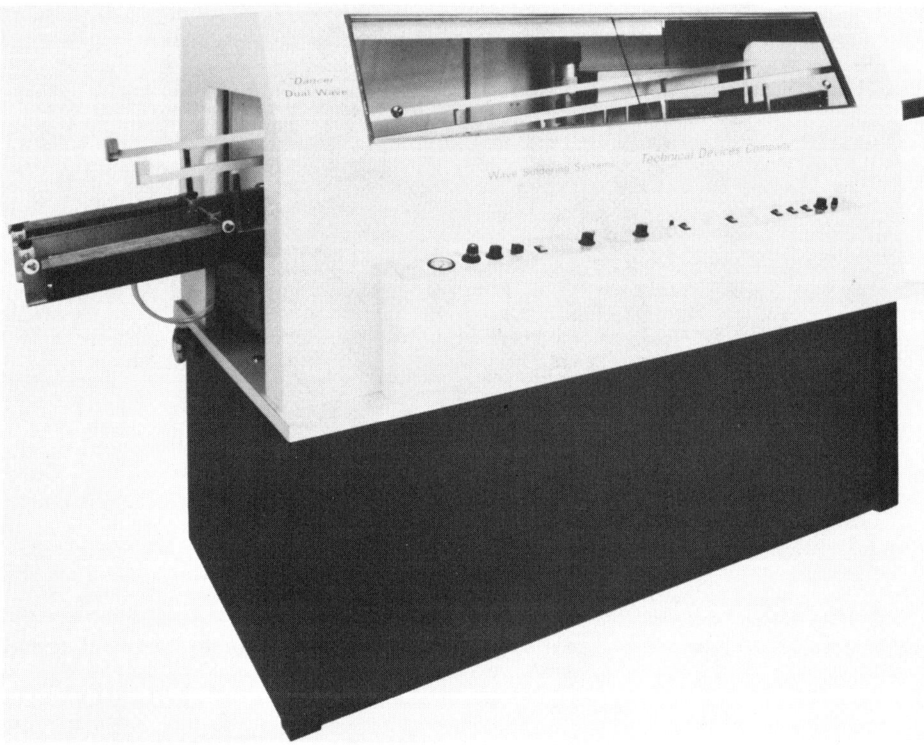

Figure 4.2 Technical Devices Mark X wave soldering system. (Courtesy of Technical Devices Company.)

- Dynamic solder pots
- Inclined conveyors
- Wave, foam, or spray fluxers
- Calrod, hot air, and quartz preheaters
- Oil injection

Following are some variables of the soldering system:

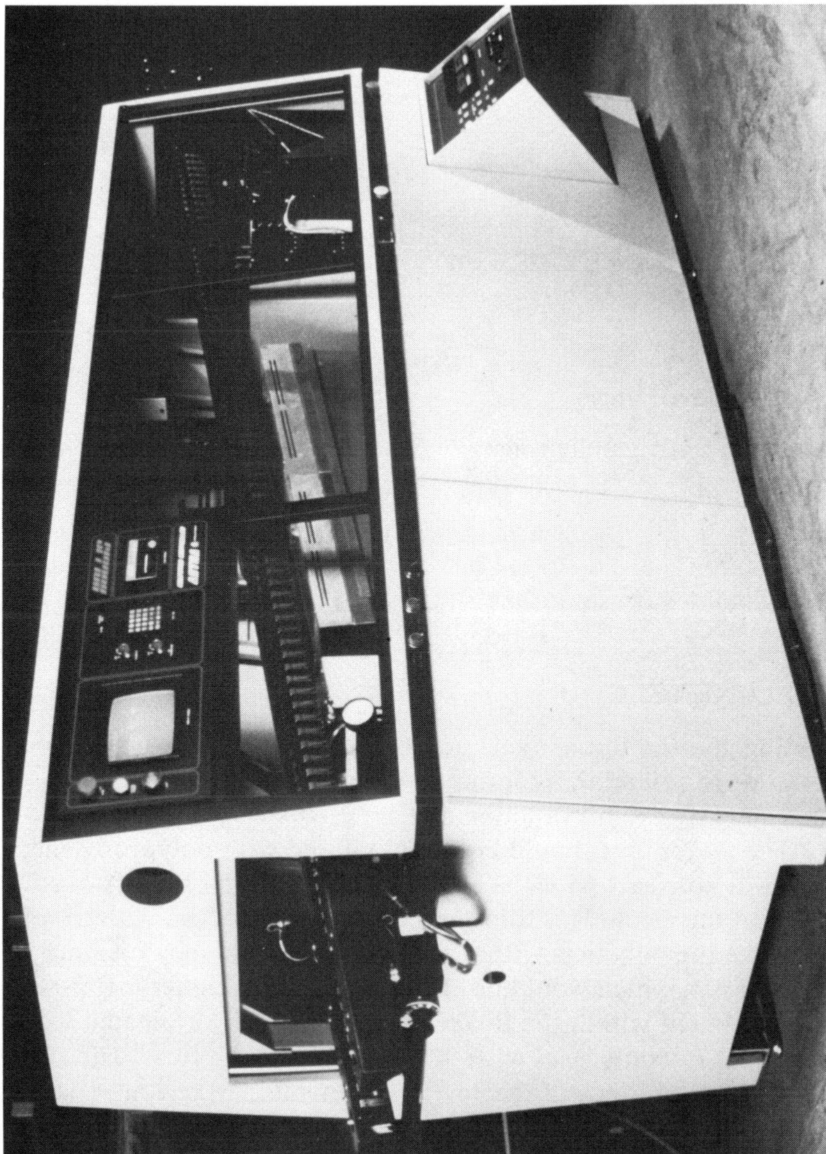

Figure 4.3 Hollis HMCS Astra wave solder system. (Courtesy of Hollis Automation Inc.)

- Flux density
- Foam head
- Wipeoff brush
- Preheaters
- Solder pot
- Solder alloy
- Solder temperature
- Conveyor
- Conveyor angle
- Conveyor parallelism
- Oil adjustment
- Dross level
- Solder wave smoothness

4.2 FLUXER SECTION

Functionally, the fluxer's sole purpose is to apply flux to the product prior to soldering. Various styles of fluxer have been designed to perform this function and are discussed below.

The fluxer unit should be manufactured with polypropylene, which will not react with the acids; however, not all plastics can withstand the alcohols within the fluxes. Alcohols can withdraw or dissolve the polymers within the plasticizers, thereby creating brittle plastic, which would not be acceptable for fluxer materials. Any metal used within the fluxer must be carefully evaluated to certify that it won't react with the activators in the flux. The acids within the flux will react with any metals and one must also be concerned with the fitting on the air stone and drain valves within the fluxer for potential deterioration. The air stone is a porous stone which is used to generate the foam head in the fluxer. As pressurized air is introduced into the stone it escapes from the stone through the pores of the stone and creates a foaming action in the flux. Reaction with metal fittings or any other metals within the fluxer will change the characteristics of the flux, and ultimately change the quality of the soldering.

Once the flux has been chosen the flux density/specific gravity must be continuously monitored. This can be measured either manually with a hydrometer or automatically with a specialized viscosity measurement system, as shown in Fig. 4.4. Automatic flux controllers perform the dual function of monitoring the viscosity and adjusting the flux to recommended specific gravities. For each viscosity a set time is identified and recorded: as the system is then activated, any variation is adjusted by the addition of either flux or thinner. Flux controllers, however, are not necessarily temperature-compensating: therefore, it is important to maintain temperature stability of the flux solution. The best way to accomplish this is to store and monitor the material outside of the wave solder system. This storage of flux outside of the wave solder system also has the advantage of continously circulating the flux through the fluxer. This prevents flux separation or stratification and enhances the effectivity of the flux.

4.2.1 Adjusting Flux Height to the Printed Wiring Assembly

Depending on which fluxer is used, adjust the fluxer dispensing system so that the flux just contacts the bottom side of the printed wiring assembly. As the assembly passes over the fluxer, certify that the flux is penetrating the plated through holes and coming up through to the component side. This fluxing action is necessary to prepare the metallic surfaces of the assembly and components for soldering.

4.2.2 Wave Fluxer

Wave fluxers are primarily used for high solid content fluxes, which are by design poor foamers. The variations in specific gravity, although important for determining the activity level, are not as important with flux viscosity. Therefore wave fluxers are sufficient to dispense flux to the assemblies.

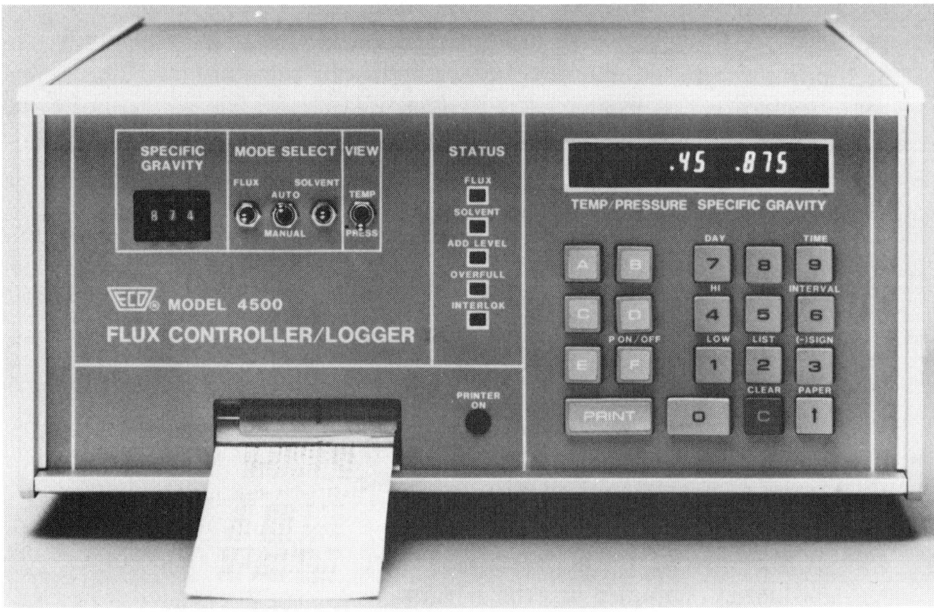

Figure 4.4 Flux density controller/logger. (Courtesy of ECD, Inc.)

The height of the flux head must be considered especially when long component leads are evident on side 2 (the solder side) of the assembly. Pumping the wave over 1/2 in. to 3/4 in. is usually standard on the types of flux system shown in Fig. 4.5.

A wave fluxer consists of the following

- Flux container of a given size to accommodate a given volume of flux
- Pump and motors to create the wave itself
- Chimney or tower to create the proper columns of flux to accommodate various component lead lengths

In automated systems, specific gravity measurement controllers monitor and control the specific gravity to certain fixed parameters.

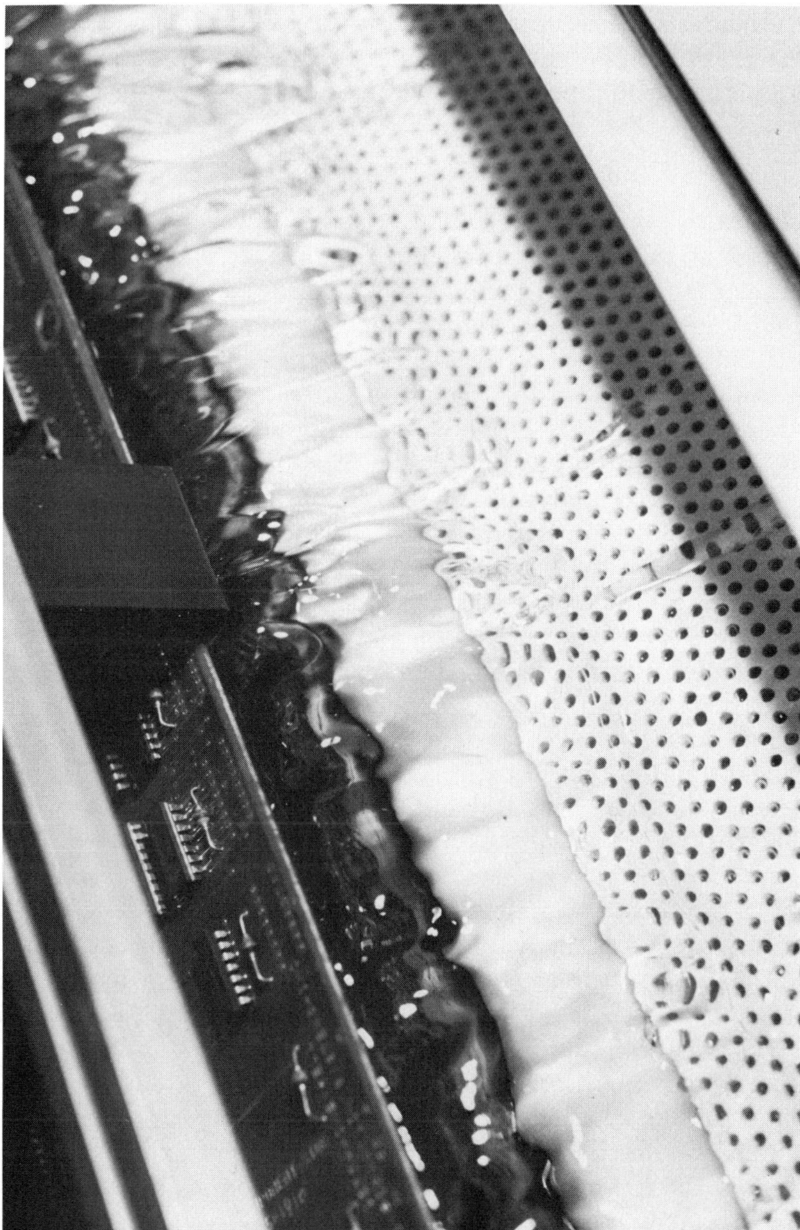

Figure 4.5 Wave fluxer. (Courtesy of Hollis Automation Inc.)

An advantage of this style fluxer is its ability to accommodate higher solid content flux and not be affected by warm or hot assemblies, as would be the case with a foam fluxer.

4.2.3 Foam Fluxer

Foam fluxers contain parts similar to those of wave fluxers except they have no pumps and motors but a perforated air stone to generate the foam head. By introducing compressed air into a perforated stone, the air is evenly distributed across the width of the fluxer and creates a foam. The chimney brushes act as an aid in increasing the height of the foam head, thereby creating support for the foam. The brushes allow the component leads to pass easily through without disturbing the components as shown in Fig. 4.6. Air pressure is utilized to control the foam head. Too much pressure will create large bubbles which are inefficient for fluxing, whereas too little pressure will not generate enough foam to flux the assembly. The air bubbles should be small and consistent to enhance the fluxing action and allow the flux to penetrate the plated through holes. The air stone should also be off the bottom surface of the fluxer; if it is placed on the bottom, entrapment areas will occur and inconsistent bubbles will be created causing discontinuities in the fluxing operation. When viewing the fluxer and the foam head, the bubbles should be small and consistent across the entire width of the fluxer.

Another element to consider is the parallelism of the conveyor to the fluxer foam head or wave head. This is to certify equal fluxing across the width of the product. Don't run any hot or warm product through the foam fluxer unit, as this will destroy the foam head.

If rosin flux is used, the liquid should be a clear amber color with the foam a light beige. It should not be dark or cloudy, which would be considered dirty and unacceptable for soldering.

Care should also be taken to certify that the flux does not come over the top of the assembly when it is coming into contact with the fluxer. Flux coming onto side 1 (component side) can become entrapped in, around, and beneath components and be difficult to remove.

Figure 4.6 Foam fluxer unit showing the chimner and control brushes. (Courtesy of Hollis Automation Inc.)

Also certify that the wipe-off brush or air knife is functioning properly and wiping the dripping flux globules off the bottom surface of the assembly. If a wipe-off brush is used, it should not deform when the pwa passes over it; if it does, then it must be lowered. If the brush is in fact being displaced, it will remove the flux from the plated through holes and potentially create defective joints.

Contamination of the flux will also prevent a consistent foam head and result in poor soldering. Similarly, lack of foam can be attributed to poor specific gravity adjustment because when the flux gets too thick it doesn't foam. Contaminants also come from a variety of other sources, the major sources being the pwbs themselves and their oxides and particulate material, the conveyor (finger type) on the return from the solder pot. These are rather obvious; however, one contaminant not as evident and which results in poor soldering quality is water. Water can come from a variety of sources, but the most prevalent sources are condensate from the air compressor and humidity in the atmosphere. Unless the plant or facility is temperature- and humidity-controlled, there is very little that can be done to prevent the water source from existing. However, the air line condensate is easily removed with a water separator installed in the air line. The only maintenance needed with this system is to drain the water on a regular basis. Another major cause of foam head deterioration is the introduction of silicon into the fluxer. Silicon is typically introduced by the thermal compound used on power components. Care must be exercised to prevent the silicon from contaminating the fluxer.

4.2.4 Spray Fluxer

Spray fluxers consist of a pressurized flux system and predetermined, configured spray nozzles. As the printed wiring assembly is conveyed into the fluxer area, sensors detect the presence and activate the pressurized system.

It is important to get spray nozzles which will cover the complete area of the largest possible assembly for the system. The problems are found when smaller assemblies are processed. Modifications are necessary to prevent excess flux from being sprayed by nozzles that are not in the region of the assembly. This extra spraying causes the flux mist to be deposited over the component side and again this flux is difficult to remove.

As the assembly is conveyorized over the fluxer area, sensors activate the fluxer nozzles and a given quantity of flux is sprayed onto the bottom side of the module. Spray fluxers are one of the most easily controlled flux applicators. The quantity of flux applied is always constant due to the amount of pressure in the line and the size of the nozzles. Overspray is going to occur, and the enclosure must be capable of withstanding the chemicals from the fluxer. Particular areas of the assembly should be protected from flux as necessary as overspray could cause some problems. Overspray onto adjacent equipment sections (i.e., conveyor, interior of cabinet) is a major concern if water-soluble flux is used. Water-soluble flux is acidic and can affect any metal components with which it comes into contact.

4.2.5 Brush Fluxer

Brush fluxers can be designed to use either a stationary or roller type of applicator. Stationary brush fluxers utilize bristle or nylon brushes which are soaking in a flux container and as the pwa passes over the brushes the flux is applied to the bottom side of the assembly. This type of flux applicator is not very consistent in applying flux, and missed areas are common.

Roller fluxers are usually not used in mass soldering operations except in board-coating applications. Any assembly having components with long leads could not be processed through a standard-type roller fluxer. Usually, roller fluxers are used in applying organic protective coatings to printed wiring boards to protect the surfaces from oxidizing and prepare them for soldering. Roller applicators are designed as a motorized roller brush unit sitting in a flux pot. As the board is conveyed over the system, the roller is activated and rotates against the direction of the board travel, applying flux to the bottom surface of the board.

There are exceptions, however, and in mass soldering systems sometimes a roller fluxer is used in a solder cut system during the secondary fluxing operation. Solder cut systems are mass soldering units with two solder pots, and an automatic lead cutter or trimmer between the two pots. The first solder pot performs the initial soldering and secures the components in the assembly. At this point, with the leads secured the assembly is conveyed over the lead cutter, where all the long leads are cut to a predetermined length. Sometimes this operation fractures the solder joints; therefore a secondary solder pot resolders the assembly repairing the fractured solder joints and also soldering the ends of the cut leads to prevent corrosion from attacking the base material of the component leads. The residual heat of the first soldering system prevents the foaming action of the standard foam fluxer system, and to properly flux the pwa for the second stage soldering roller brush fluxers are sometimes used.

4.3 PREHEAT

This section is usually one of the elements of the solder system kept as a variable. The reason is the variability of each assembly being soldered.

Preheating, the second stage of a mass soldering process, utilizes one of the following types of heat source: cal rod heaters, quartz lamps, hot air blowers, or hot plates to prepare the assembly for soldering. See Figs. 4.7-4.9 for various styles of preheaters.

The necessity of preheat must be considered. Soldering is a process performed above a certain temperature for a given time period. Therefore instead of letting the massive quantity of molten solder raise the pwa temperature to the soldering temperatures, preheaters are used to elevate the ambient temperature of the pwa and reduce the time necessary for soldering. In effect, preheaters allow the increase in conveyor speed.

If no preheaters were used, all the anecdotal stories regarding flux volatiles drying, preparation to reduce thermal shock, activa-

Figure 4.7 Calrod preheaters. (Courtesy of Hollis Automation Inc.)

Figure 4.8 Quartz preheaters. (Courtesy of Hollis Automation Inc.)

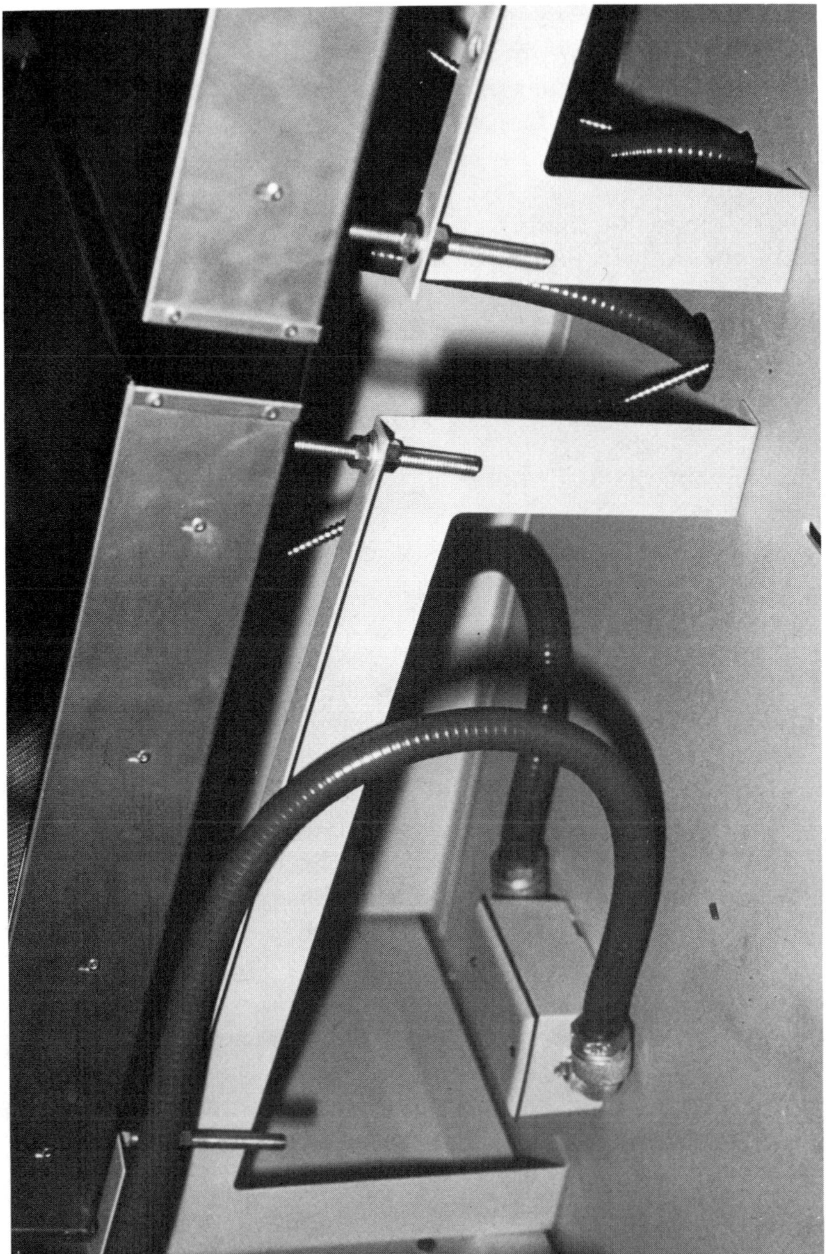

Figure 4.9 Hot plate preheaters. (Courtesy of Hollis Automation Inc.)

tion of the flux, etc., would be accomplished when the assembly contacted the solder wave. Boards are qualified to withstand 560°F for 10 sec per military standards. Albeit not all commercial assemblies are qualified as such, they should still be able to withstand a temperature soak on molten solder for a given time period without being destroyed or deteriorated.

The phrase "preparing a board for soldering" hints of ambiguity due to the following elements:

- It implies the activators within the flux will be brought to their maximum activity levels,
- The alcohol and volatiles vehicles will be driven off the surfaces of the assembly, leaving only the fluxing material to provide the oxide removal.

The alcohols have low vapor pressures; therefore they evaporate quickly under normal exposure to air. However, under exposure to accelerated heat, the evaporation is much quicker. The quick evaporation creates increased vapor pressure within the confines of the preheaters, and since the fumes are volatile, explosions and fires are common if proper safety precautions are not taken.

The reduction or elimination of thermal shock by preheating is one issue that is highly debated in soldering circles. Board laminate material FR-4 per MIL-P-55110 is supposed to withstand thermal shock of 10 sec at 560°F. Soldering process temperature parameters are not as severe as those required per MIL-P-55110. The implication is that preheat may not be necessary; the boards should be able to take the shock. Then why does preheat exist? It exists for all the reasons previously mentioned; however, the major reason for preheat is to develop the ability to increase the conveyor speed.

The design of preheaters incorporates reflector plates to increase the efficiency of the cal rod or quartz preheaters. It is important they be kept clean so that proper reflective characteristics can be utilized to refocus the energy toward the product being conveyorized.

If hot plate preheaters are used they must also be kept clean. Flux residues which drip onto the various types of preheaters must be continually removed to enhance the effectivity of the preheaters. Flux contaminants are almost classified as insulating types of materials; therefore, it not removed, their insulating properties will prevent proper preheat.

Hot air preheaters are primarily used to remove the volatiles in the flux. Precautions are usually taken to prevent the air flow from affecting the evaporation rate of the alcohol in the fluxes.

Again, preheaters allow for increased conveyor speed. Therefore, to what temperature should the preheaters bring the assemblies must be answered in developing the process.

4.3.1 How to Set the Preheaters

Industrial practices seem to demand 200–215°F on the top surface of the printed wiring assembly. This is accomplished by utilizing convection heat from the preheater source to heat the bottom side of the assembly. The heat then conducts through the assembly and heats the top side. This is accomplished by the ability of the base laminate to act as a thermal conductor when in fact it is a glass insulator.

The ease with which heat transfers through the laminate makes the preheaters appear all the more efficient. However, as the laminate or printed wiring assemblies get thicker, i.e., greater than 0.090 in., or are built with more inner layers, e.g., multilayer assemblies, the transfer of heat through the assembly becomes more difficult and at times almost impossible due to the insulation properties. To overcome this phenomenon, top side preheaters are utilized, or quartz-type units are used to heat the oven chamber. The concept is to convert the preheat section into an oven; however, the inefficiency of convection heat, to get product hot quickly, is still suspect in theory. Long preheat sections therefore are designed to get the maximum heat transfer around and onto all the surfaces of the product.

Also, when all the heat is applied to the bottom side of the assembly, and the product is a multilayer type and the inner layers act as heat sinks, the first or bottom layer heats up at a different rate than the others and behaves like a bimetal strip and the assembly develops excessive warpage characteristics.

It is recommended to attach thermocouples to both the top and bottom sides of the assembly and measure the exact temperatures as the product is being transported through the system. Making thermal profiles of each type of assembly being soldered will create a library of charts which can be used to determine the proper power settings for the system preheaters. The temperature difference between side 1 (component side) and side 2 (solder side) should not exceed 75°F.

4.4 SOLDER WAVE CONFIGURATION

Over the years, equipment manufacturers have developed various solder wave configurations that supposedly are different and yet uniquely similar. Hence, when one considers the basics of soldering, the process of joining two metals using a third metal performed at some time/temperature condition, the only purpose of the wave is to get solder to the components to be joined. Therefore with this premise, there are a variety of different dynamic wave or static systems which can be used, such as the bidirectional or unidirectional solder wave concepts. Figure 4.10 shows various wave configurations.

4.4.1 Dynamic Wave Concept

All systems which utilize a pump mechanism to move the solder through a wave generator are considered dynamic wave solder systems. The implication is not only that the product is going over the wave, but that the molten metal is also moving, thereby enhancing the cleaning action of the flux. Hence if any oxides are too thick to be removed with the flux yet are on the surface of the tin-lead plating, the action of the wave during the reflow operation will mechanically remove that solder coating to produce good soldering conditions. In the case of tin or tin-lead plating, the excessive plating once in the molten condition is washed by the flow of the solder from the pot, or the conveyorized action of the conveyor. This washing action benefit is attributed only to plated types of modules where the plating has some measurable thickness, i.e., greater than 0.000050 in. Certain elements, however, can also hinder this washing action. Examples are when all the traces are copper or copper-plated, such as those found on print and etch boards, or with thin solderable coatings consisting only of copper-tin intermetallics.

4.4.2 Wave Configuration Analysis

Cross-sectional views of wave solder configurations describe the solder flow and forces associated with the dynamics of the system. The forces developed by the moving solder creates many other multidirectional forces that enhance the wetting action of the solder. These forces coupled with an inclined conveyor to transport the product create angles of solder release, where the pullback forces are greater than the adhesive forces of the solder. This phenomenon reduces and eliminates icicles and helps in developing informational data for the component layout of printed wiring boards.

4.4.3 Unidirectional Solder Pot

This wave design is configured so that the solder flows in only one direction, toward the fluxer, when no product is being soldered. When a printed wiring assembly is passed over the pumped wave, the far or back side of the solder pot is mechanically moved by the friction of the assembly going over the wave as exhibited per *Fr* in Fig. 4.11. The pull-back force in this design is generated by the movement of the printed wiring assembly while it is going over the wave. Therefore, *Fr Cos Ø* must be greater than *Fr Sin Ø* to prevent icicles from forming. See Fig. 4.12 regarding icicle formation. As can be seen, the pull-back forces created as the pwa leaves the wave are critical to prevent the creation of icicles and solder shorts. Icicles and solder shorts are two of the most common defects attributable to the wave soldering process.

As can be seen from Fig. 4.12 and the formula presented, if *Fr Sin Ø* becomes too great and wetting action is occurring, then

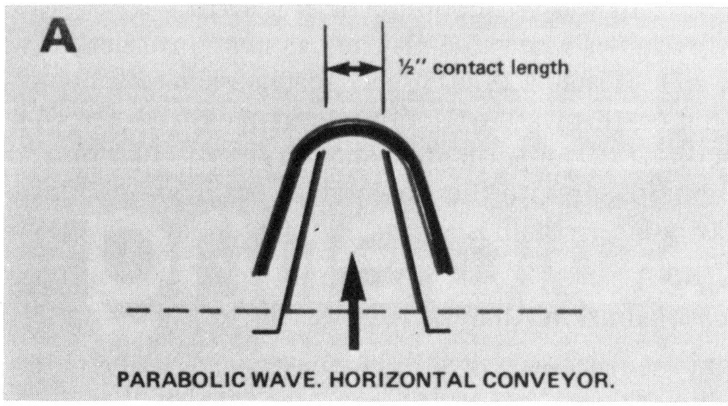

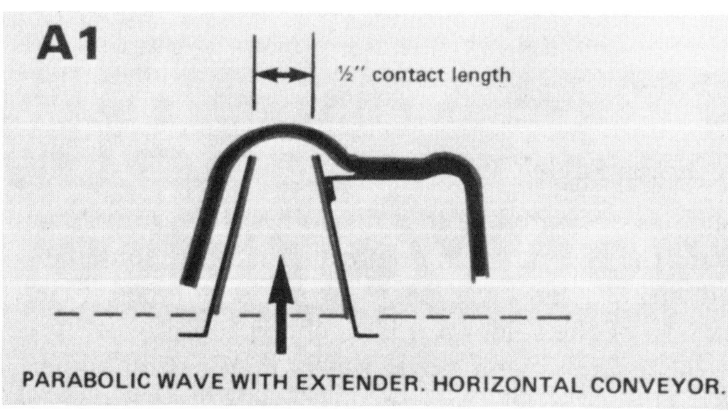

Figure 4.10 Wave generator configurations. (Courtesy of Hollis Automation Inc.)

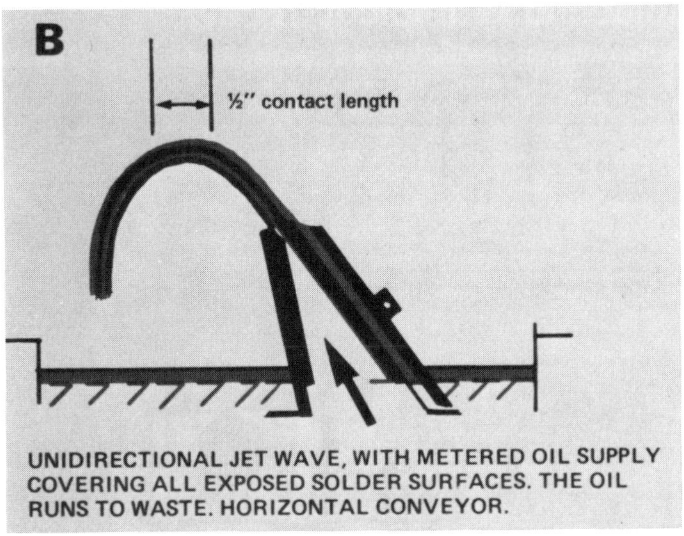

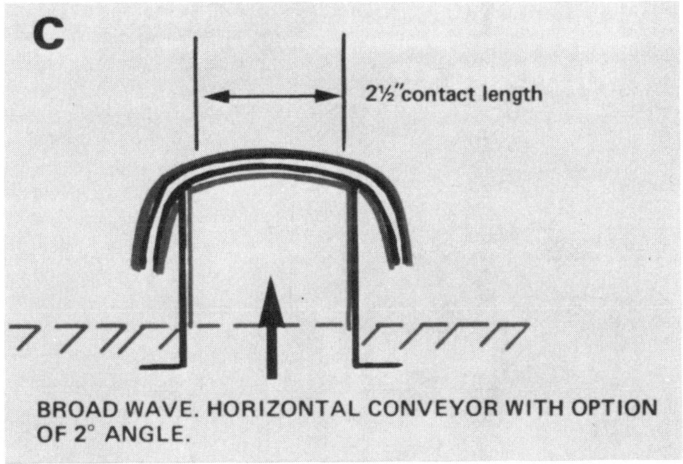

Figure 4.10 (Continued)

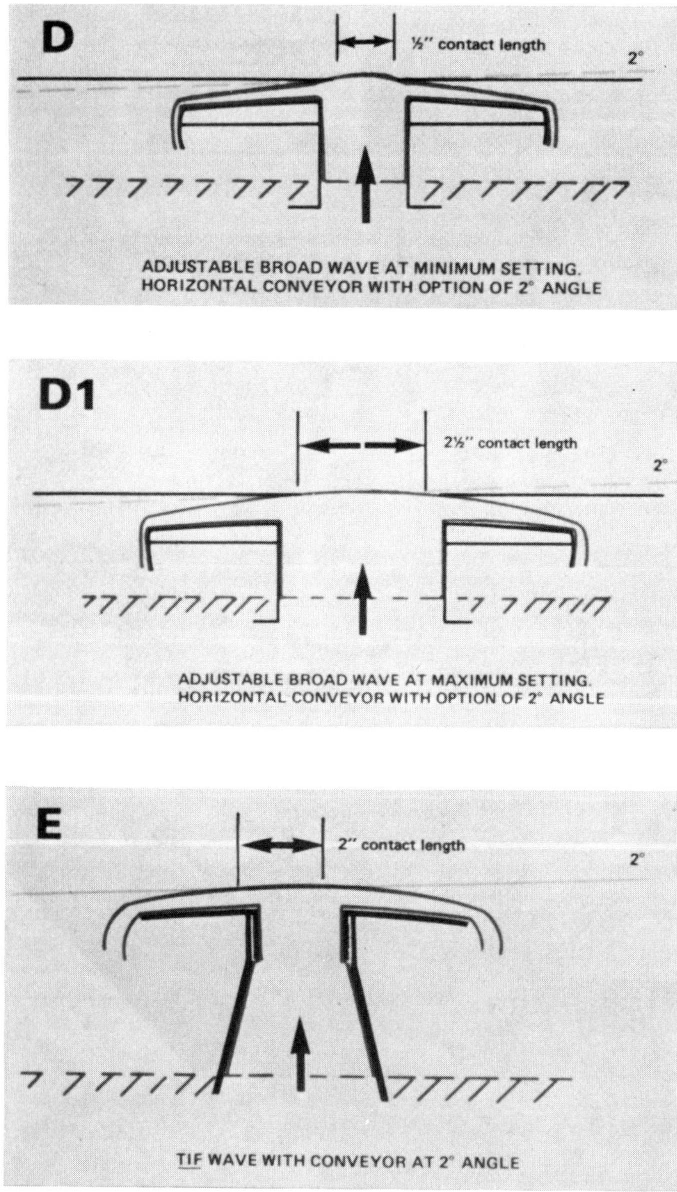

Figure 4.10 (Continued)

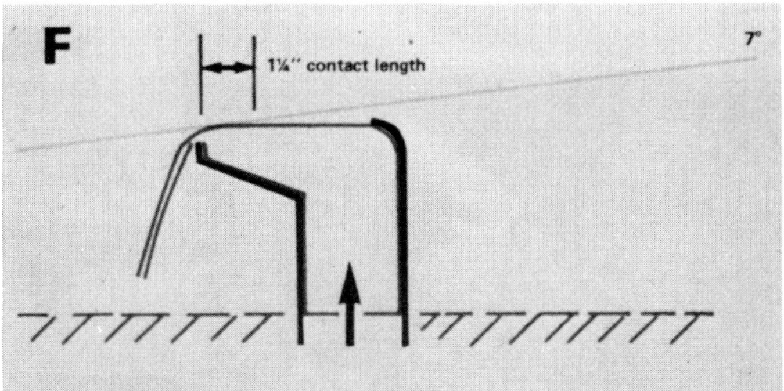

DELTA WAVE WITH CONVEYOR AT 7° ANGLE

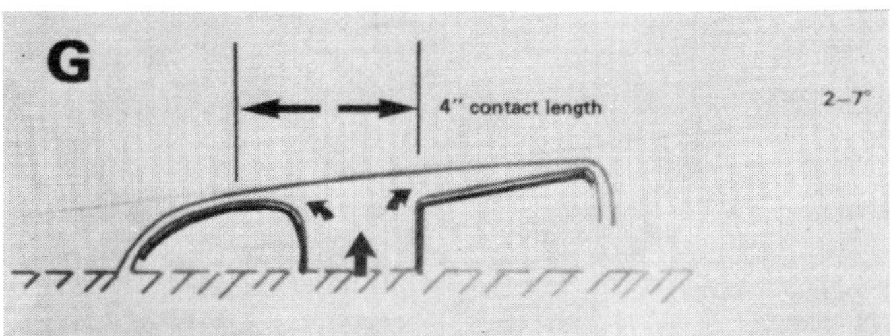

STANDARD LAMBDA WAVE. CONVEYOR ANGLE 2–7°

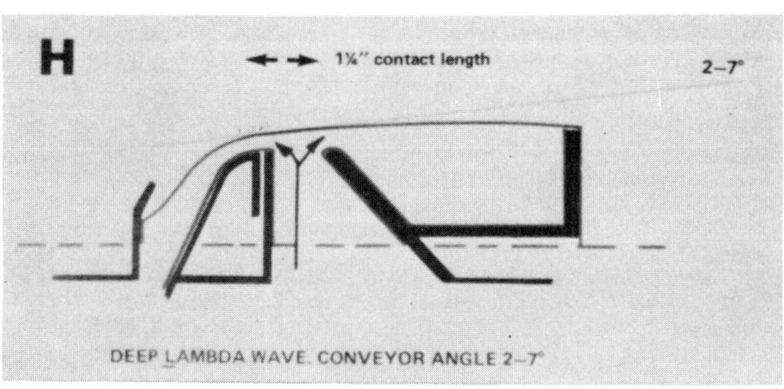

DEEP LAMBDA WAVE. CONVEYOR ANGLE 2–7°

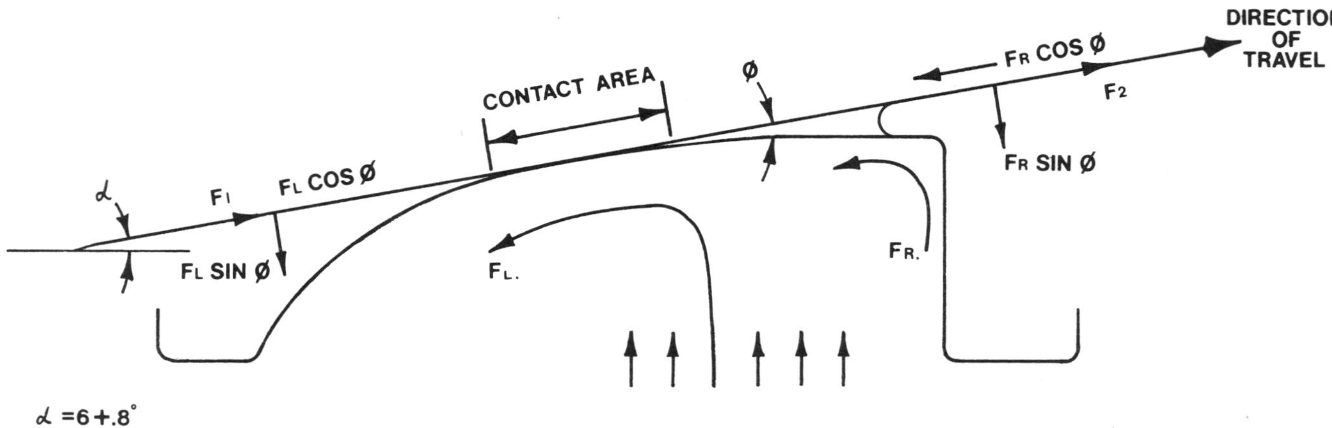

Figure 4.11 Unidirectional wave diagram.

Figure 4.12 Icicle formation. (Courtesy of Hollis Automation Inc.)

the solder will slip upon itself and withdraw from the plated through holes. This phenomenon creates shallow fillets in plated through holes with large lead-to-hole ratios. Example of the solder fall out would be with integrated circuit (IC) leads.

4.4.4 Bidirectional Solder Pot

When the webbing action is too far away or extended to position B, *Fy Sin ∅* overcomes *Fy Cos ∅*; icicles and solder shorts are thus created between component leads as shown in Fig. 4.13. If the conveyor is set at a horizontal position, the theoretical B location would exist and icicles would occur. This has been proven many times over the years, which is one of the reasons for the incline of the conveyor.

4.4.5 Cascading Wave System

Cascading wave systems were designed for soldering printed wiring assemblies as many times as there were ripples in the cascade.

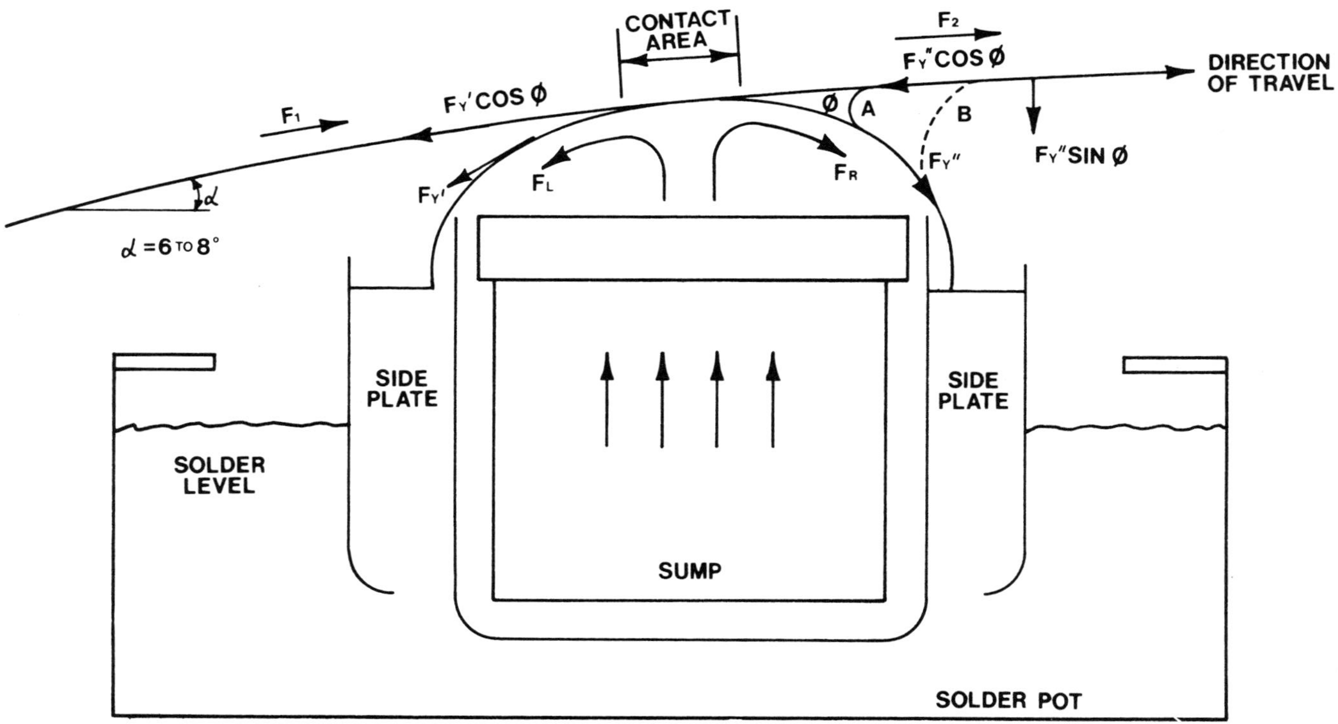

Figure 4.13 Bidirectional wave diagram.

This was done with the belief that each succeeding wave would remove the problems created by the preceding wave. Although used extensively by some companies, the idea never seemed to have caught on, and the major equipment manufacturers don't offer this configuration any longer.

4.4.6 Static Solder Pot

Initially, all mass soldering operations were performed in a static solder pot called "dip solder pots." Utilizing a pair of tongs or pliers to grasp or hold the product, the product is lowered onto the surface of the solder for a given time period until the product is soldered. Although this process had its limitation with various product sizes and environmental and employee hazards, the process was more efficient than the manual soldering operation where each joint was individually soldered.

Presently static solder pots are still used for pretinning operations. Any mass soldering performed with static solder pots are usually heavily conveyorized and sophisticated. The Zevatron™ * mass soldering system is such a system. Some advantages of static pots are the small quantity of solder in the pots, the lack of moving parts, such as motors and pumps, and the relatively low dross generation. However, since the solder is not moving, some mechanism must still be used to remove the dross or oxide film buildup that occurs on the solder surface prior to conveyorizing the assembly over the solder pot. The conveyor and fixtures must also be designed for the product to enter and leave the solder pot at an angle. This is to stimulate a greater force by the molten metal to prevent icicles and shorts. These issues were the major contributors for the development and creation of dynamic wave solder systems. Various style solder pots are shown in Figs. 4.14 and 4.15.

4.4.7 Setting Solder Pot Temperature

The solder pot temperature is a constant; once set it is not changed. Typical settings are anywhere from 480°F to 515°F.

---

*Zeva Electric Corporation.

Figure 4.14 Wave generator and pump assembly. (Courtesy of Hollis Automation Inc.)

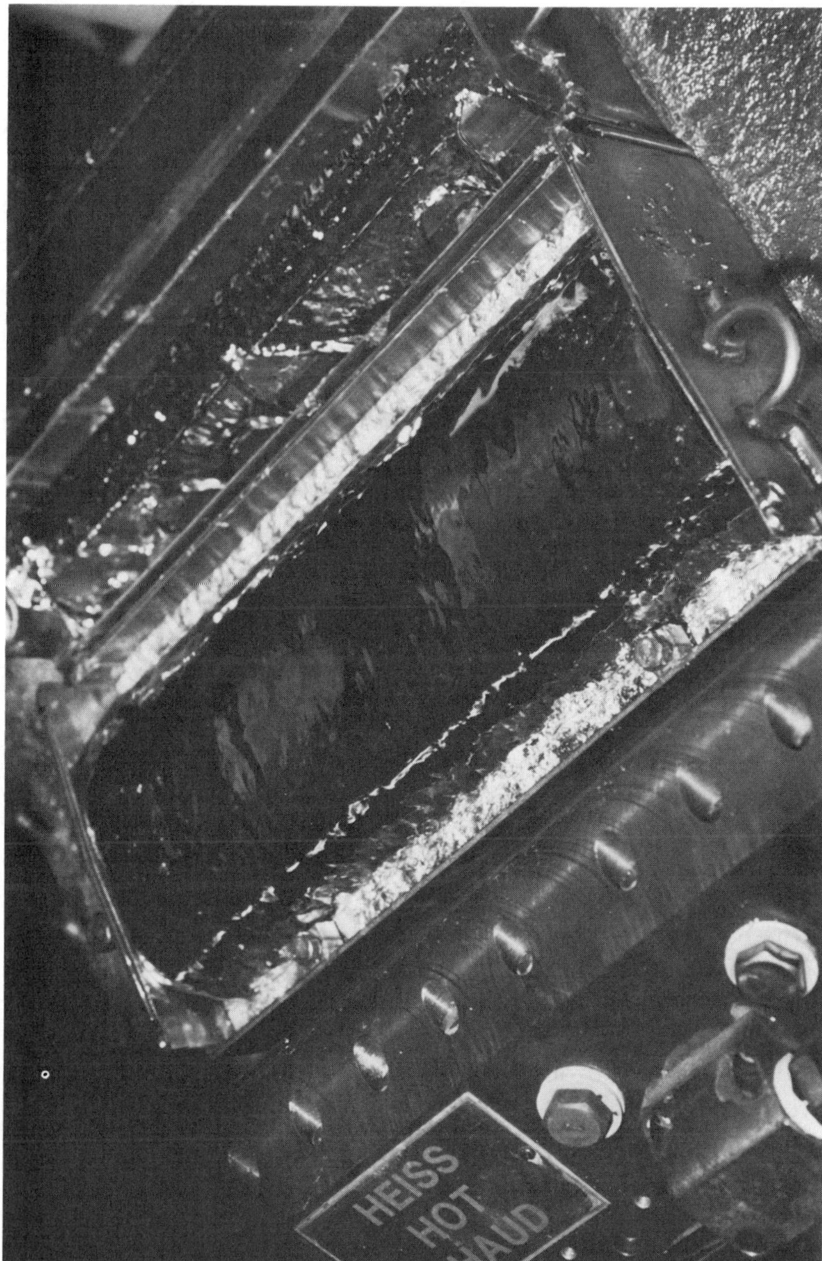

Figure 4.15 Chip Z wave and hot air knife system. (Courtesy of Hollis Automation Inc.)

Electronic controllers such as that in Fig. 4.16 are used to monitor the upper and lower temperatures. Charts such as Fig. 4.17 exhibit the monitoring data of the system. The temperature range chosen for any wave-soldering system according to all soldering texts available is 100–125°F above the melting point of the alloy used. The reasoning behind this high temperature is to accommodate or compensate for the heat loss during pumping of the solder up to the top of the wave, and the temperature loss by the heat sinking effect of the assemblies as they are being soldered.

The temperature controllers are employed to prevent high-temperature runaway conditions from occurring and low end temperature shutdown conditions. Too high a temperature will delaminate the assembly and potentially cause the oil to catch on fire, if in fact oil is used. The low end is to prevent pump and motor damage; hence, if the temperature gets too low, the controllers deactivate the power to the pump motor. The low end temperature controllers are used to monitor the heaters within the solder pot. For example, if a heater coil does not function, it will affect the overall solder pot temperature and the system will shut down.

When soldering, a measure of good soldering is how the solder flows onto the pads on side 2 or up into plated through holes. However, to accomplish this smooth flow of solder on the metallic surface a combination of preheat and time in the molten solder is required. With the proper heat being applied, the solder will flow smoothly with no interruptions, either onto the pads or into the plated through holes.

When soldering multilayer assemblies if the solder does not flow up into the plated through holes, more preheat or longer times in the solder are necessary. The only way the solder will rise into the plated through holes is if the plated through hole itself reaches soldering temperatures.

4.5.8 Establishing Soldering Parameters

Hence, when establishing the soldering parameters the important issues to consider are:

Figure 4.16 Flowtrack controller. (Courtesy of ECD Inc.)

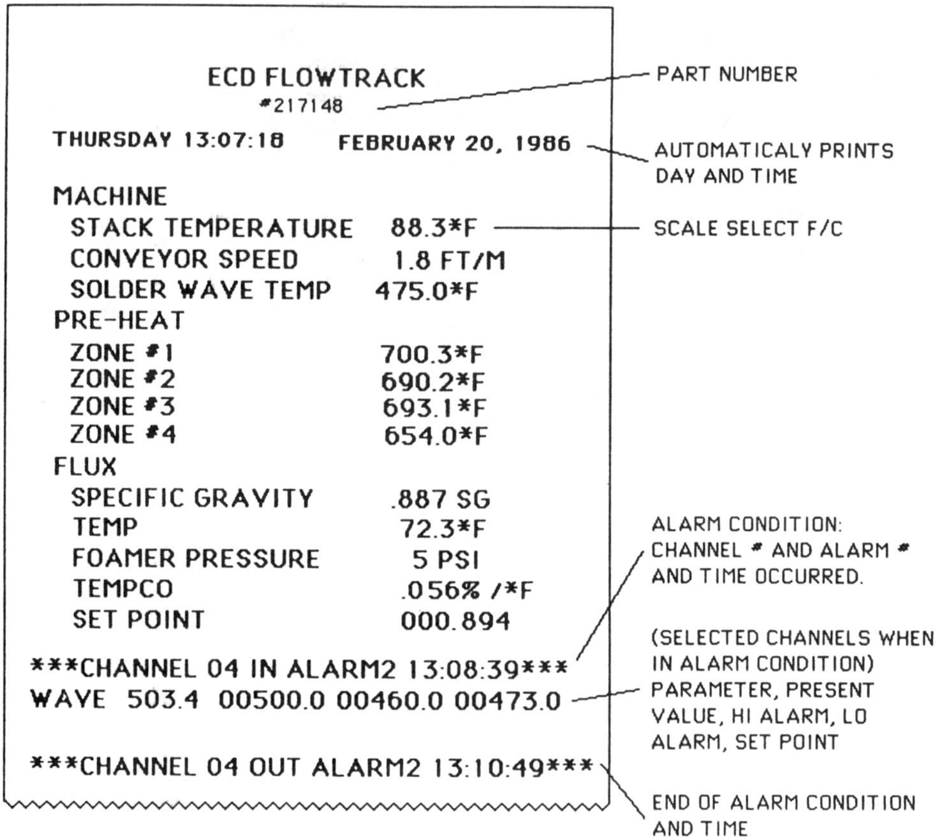

Figure 4.17 Flowtrack printout. (Courtesy of ECD Inc.)

- Time in the wave
- Speed through the wave
- Temperature of the wave
- Oil intermix
- Flow of solder from dynamics of the system

Time in wave is measured or established in a variety of ways, some of which are anecdotal and other scientific.

Anecdotal methods are to run the assembly by submerging half of its thickness in the wave. This was originally used as a premise that the hydraulic pressure of the wave would mechanically push the solder up into each and every plated through hole. Adjusting the wave so the assembly is half its depth in the wave is probably one of the most difficult elements to achieve. Submerging the boards 0.029, 0.032, and 0.041 in., etc., is subjective at best, and in addition solder plows or dams are used to prevent solder from coming over the top of the assembly, to prevent warpage or twisting during the soldering operation.

The only measure to use is the contact area of the pwa to the surface of the wave. This is accomplished using a Lev-Chek* (see Fig. 4.18) thermal glass plate. As the Lev-Chek plate passes over the wave, the contact width with the molten solder can be measured. Correlating this contact width with the conveyor speed, the time in solder can be determined. From best industrial experience, the ultimate soldering time is a minimum of 1 sec to a maximum of 3 sec. A method of determining time in the solder is as follows:

> Conveyor speed of 5 ft/min, converted to in./min = 60 in./min or 1 in./sec. Therefore, if contact length is 2 in. and conveyor speed is 5 ft/min, then the time inthe solder is 2 sec.

When using the Lev-Chek it is also important to certify that the conveyor is parallel to the width of the solder pot. This is checked by observing the contact length on both the near and far sides and making sure they are equal. This applies whether a fixture conveyor or finger conveyor is used on the soldering system.

The solder falloff from the wave area must be kep to a minimum, the intent being to keep the quantity of solder exposed to the environment as minimal as possible. Also, if side plates are used they should be properly adjusted: excessive turbulence is created if they aren't, and more dross forms on the surface of the

---

*Hexacon Electric Co., trade name.

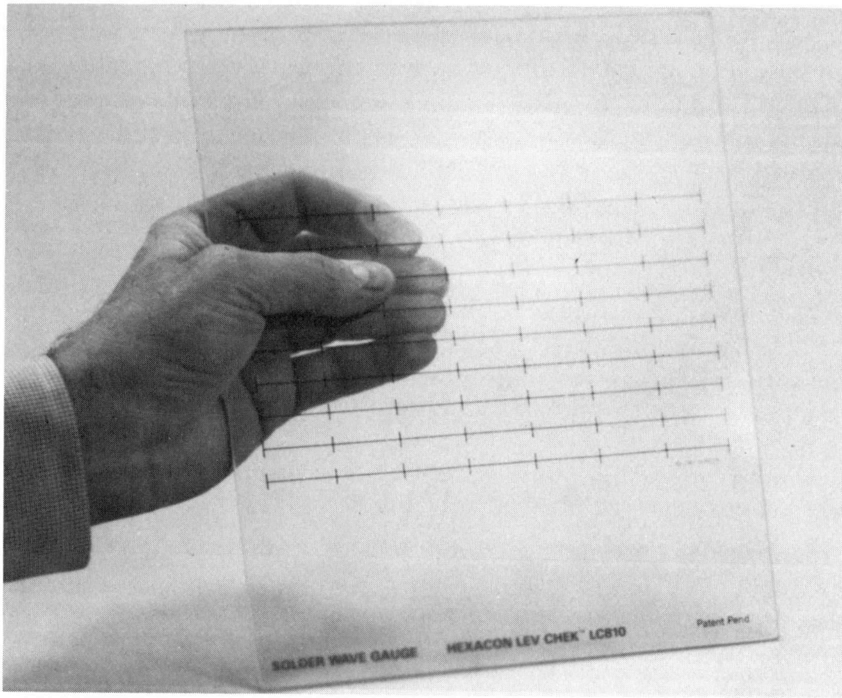

Figure 4.18 Lev-Chek solder wave gauge. Lev-Chek solder wave gauges enable easy and accurate adjustments of automatic wave-soldering machines. (Courtesy of Hexacon Electric Co., P. O. Box 36, 161 West Clay Avenue, Roselle Park, N. J. 07204.)

pot. Dross is the oxided byproduct of the molten tin-lead. It includes nonmetallic byproducts which are integral to the solder and metallic impurities from the soldering operation, such as copper from the base laminate, aluminum from fixtures, and the like. Handling of dross must be performed under controlled conditions to prevent lead contamination.

Dross is a powdery substance containing lead particles that become airborne if not handled properly. Therefore, safety breather masks and gloves must be used when handling and discarding dross. A certain amount of dross can be reclaimed and this is accomplished by selling the dross to the vendor supplying manufacturing solder materials.

If oil is used, it becomes a hazardous material once used. The oil material has both dissovled and suspended solids in its constituency. This inbred material has lead, which is poisonous to the environment. Therefore, spent oil must be handled as a hazardous waste, and proper precautions must be taken for disposal.

4.5 CONVEYOR

The conveyor is the most important mechanism of the system. It transports the product across all the sections of the systems. There are two styles of conveyors:

- Finger type
- Fixture type

In a finger-type conveyor, such as that shown in Fig. 4.19, the assemblies are placed directly into the fingers which hold and move them through the system. Fixture conveyors utilize fixtures to support and protect certain areas of the assemblies through the system.

Printed wiring board expansion during soldering is a known phenomenon. Expansion of 0.040 in. is not uncommon for assemblies that are 12-16 in. in width, and this number could be larger if wider product is soldered. To compensate for this expansion, the conveyor rails are adjusted in a nonparallel condition with the wider end at the exit end of the conveyor to accommodate the expansion.

This adjustment is necessary. If the conveyor is adjusted to be parallel, then as expansion occurs, the product will have no room for expansion and will bow, potentially allowing the solder to flow up into the plated through holes and onto the component side. Depending on the configuration of the artwork on side 1, solder shorts will result.

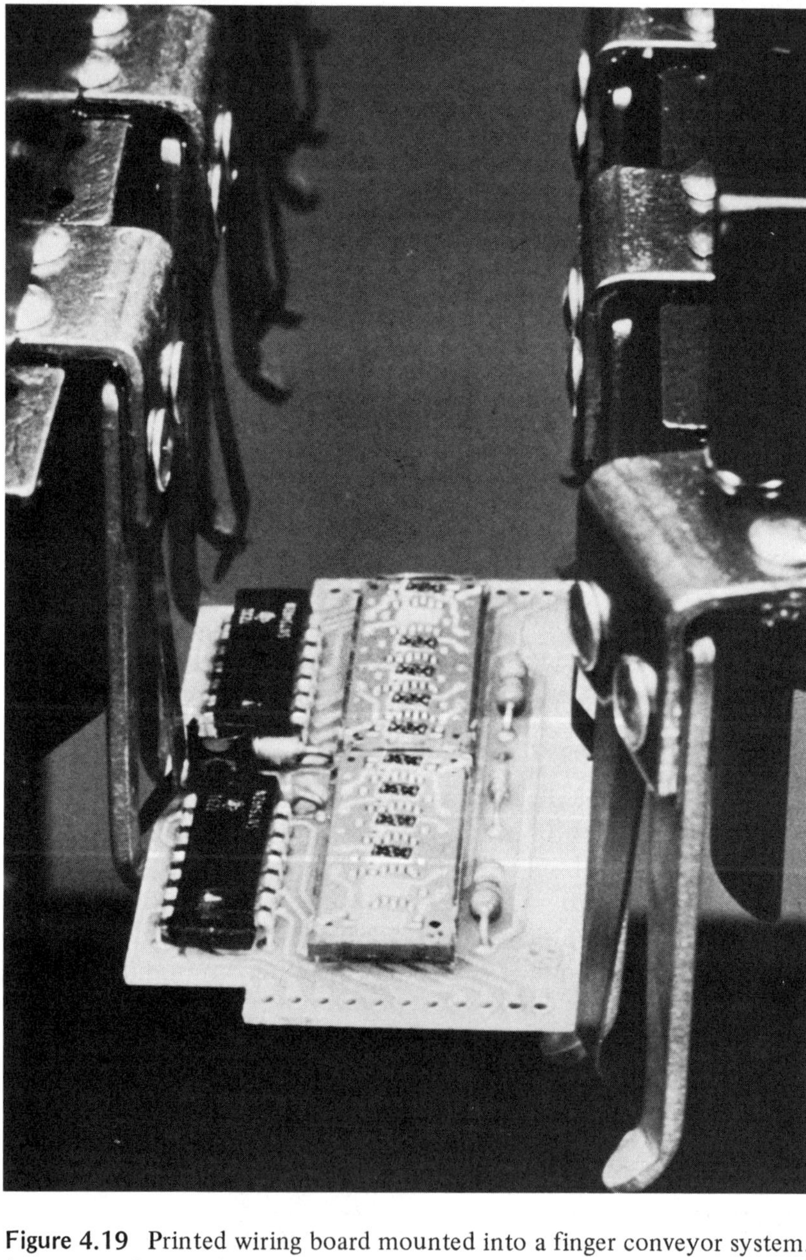

Figure 4.19 Printed wiring board mounted into a finger conveyor system. (Courtesy of Hollis Automation Inc.)

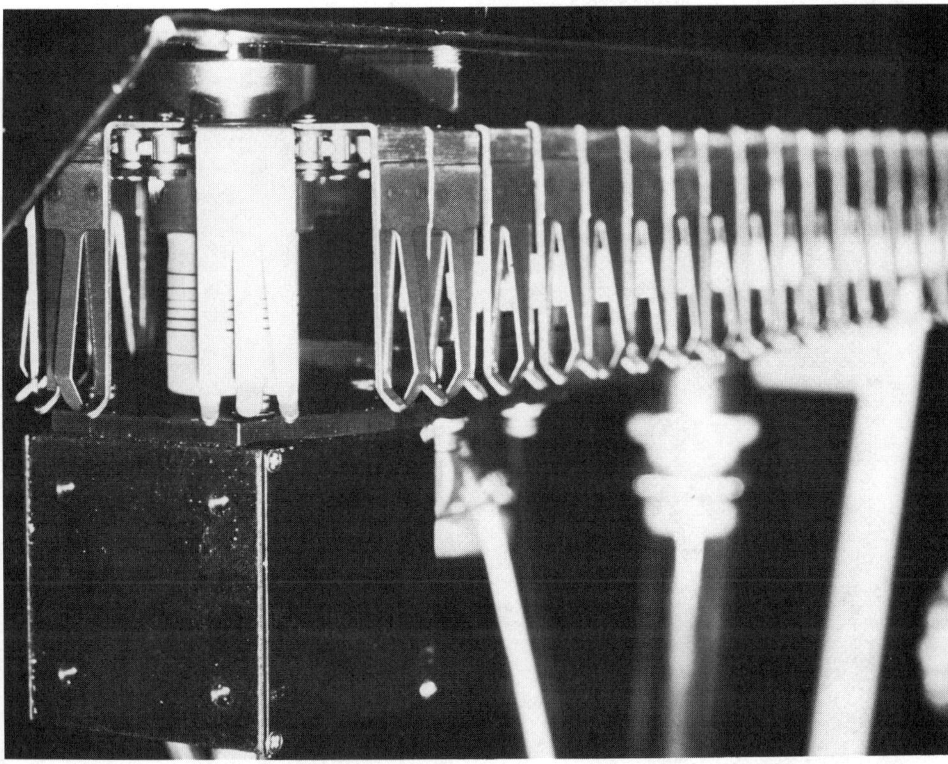

Figure 4.20 Conveyor fingers assembled to drive chain. (Courtesy of Hollis Automation Inc.)

4.5.1 Finger Conveyor

In finger conveyors, metallic fingers are attached to the drive chains to hold the assemblies in place as shown in Fig. 4.20. The configuration of the fingers vary depending on the preference of the user. In all cases, the fingers are designed to hold the assemblies. Some are configured for snap-in, others are used with automatic feeders, and others are designed for heavy modules or fixtures. Note right angle fingers in Fig. 4.21.

Finger conveyors are used when product sizes are standardized. With this type of conveyor, only one size of pwa can be

Figure 4.21 Conveyor with right-angle fingers for handling heavy assemblies. Note adjustment wheel on right hand rail. (Courtesy of Hollis Automation Inc.)

soldered at any one time. This is a disadvantage that prevents the soldering of different assembly types consecutively over the solder process. Wave solder systems do not react fast enough to compensate for these various assembly types processed consecutively on the conveyor. If various preheat and conveyor speeds are necessary to achieve maximum quality results, running different sizes consecutively will have a definite negative impact on the quality of the modules.

Important features of conveyors are the following:

- Finger material
- Conveyor rails
  - Materials
  - Construction
- Drive mechanization
- Ease of operation

In deciding which finger configuration is best for any application, consider:

- Board thickness
- Protruding components along the edge of the assembly
- Length and width of the assembly
- Gold-plated fingers
- The need for a stiffener to prevent warpage
- Component leads too close to the edge of the assembly
- Board expansion
- Board security
- Board warpage

Security implies the ability of the finger to hold the assembly as it is being conveyed through the system. The portion of the finger on which the assembly rests should be designed and chosen considering the maximum weight of the product and the closeness of the components to the edge of the assembly. As an example with the single-part finger, board thickness impacts the location of the seating plane in the finger as shown in Fig. 4.22.

The solder side of the module is at a different level for the various board thicknesses: hence, without any conveyor height adjustment the assemblies will run at a different depth in the wave. Also with thinner boards, they are positioned deeper in the slot of the finger; and if components are mounted along the edge of the assembly, the potential for missing solders or nonsoldered component leads exists, especially if leaded components are near the fingers. The fingers act as heat sinks and prevents the modules from reaching proper soldering temperatures.

Another aspect of the fingers that must be considered is the forces applied by the finger to overcome the forces applied by the wave. The forces applied by the finger must be greater than the forces of the wave or the assembly will be pushed up away from the fingers and this will impact the soldering time and contact spot of the wave (see Fig. 4.23).

Standard types of finger conveyors are designed for 0.062-ft-thick assemblies. With this thickness the spring clip applies enough forces to keep the assemblies in the fingers. Again, thicker boards prevent this spring clip from applying downward forces onto the module, thereby allowing the module to be pushed up into the fingers and reducing the contact area with the wave.

Hence, when choosing the proper finger configuration (several are shown in Fig. 4.24), take into consideration the board thickness and make sure the assembly is secure in the fingers throughout the entire length of the conveyors. Following these criteria will prevent the assembly from moving during the soldering process.

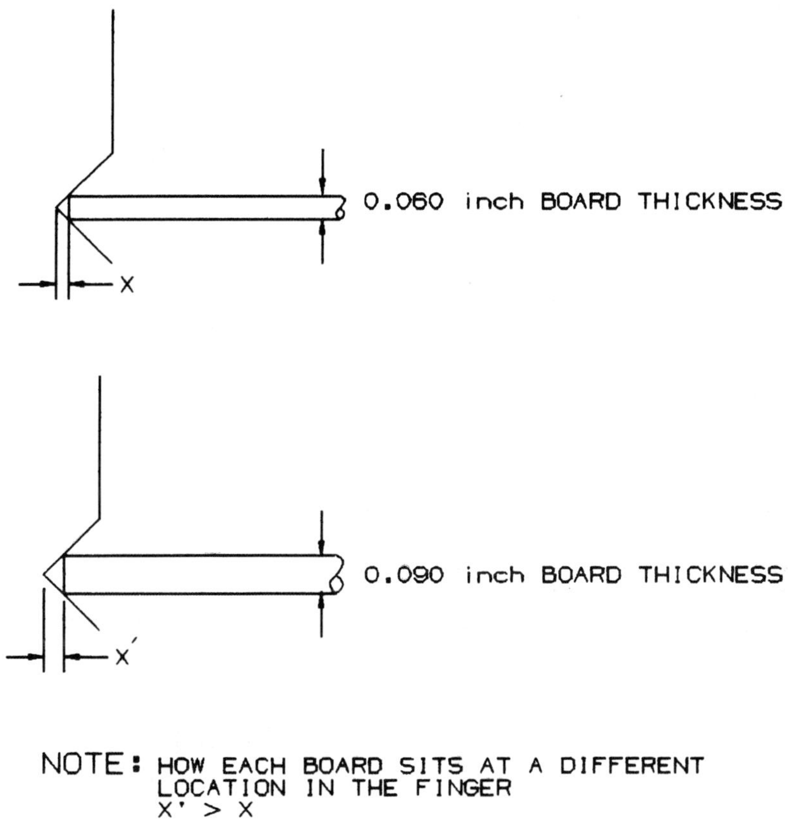

Figure 4.22 Seating plane of various thickness boards in a solid-type finger configuration.

4.5.2 Conveyor Construction

All materials used must be impervious to the chemicals of the process, i.e., fluxes, solder pot oil, and solder. Fingers must be made of materials that will not solder, such as titanium. The fingers must be designed to grasp and not damage the boards, therefore, some mechanism must be in place to hold the product be it either a board or a fixture. Conveyor rails must be designed so that the heat generated by the preheaters and solder pot will not

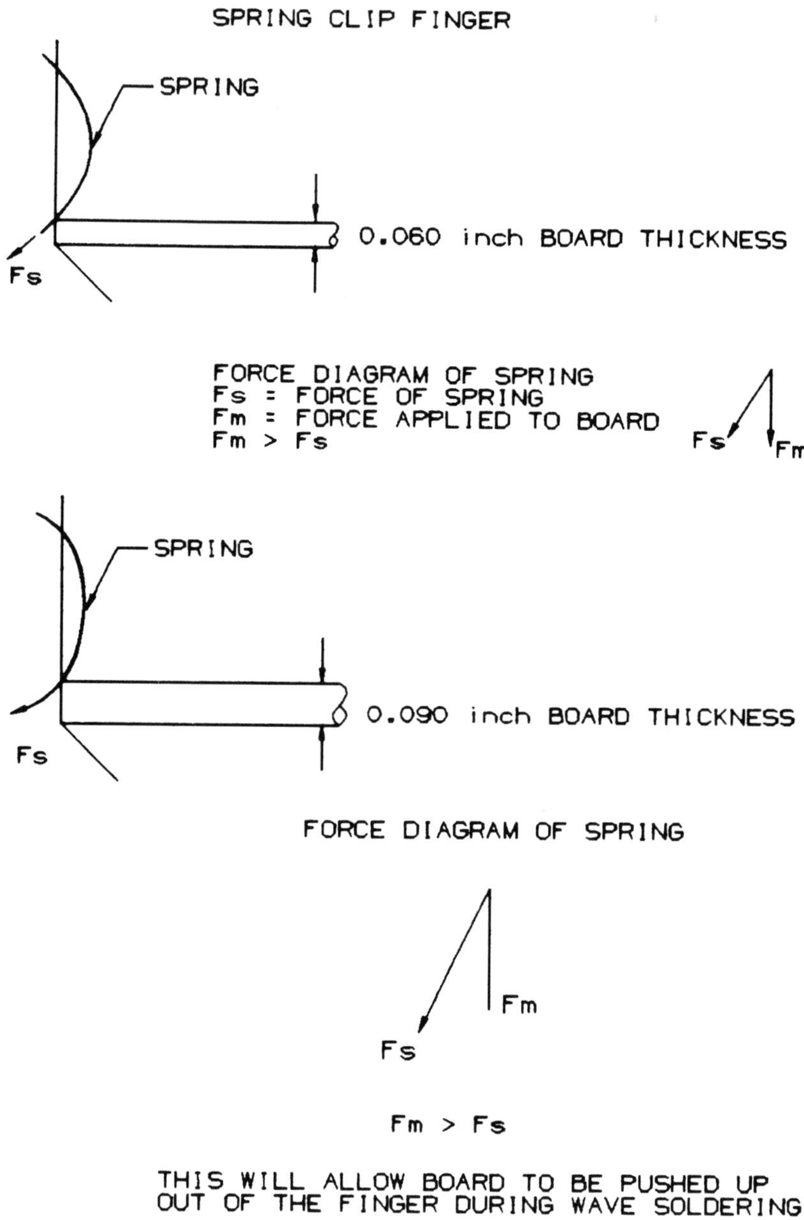

Figure 4.23 Seating plane of various thickness boards in a spring-type finger.

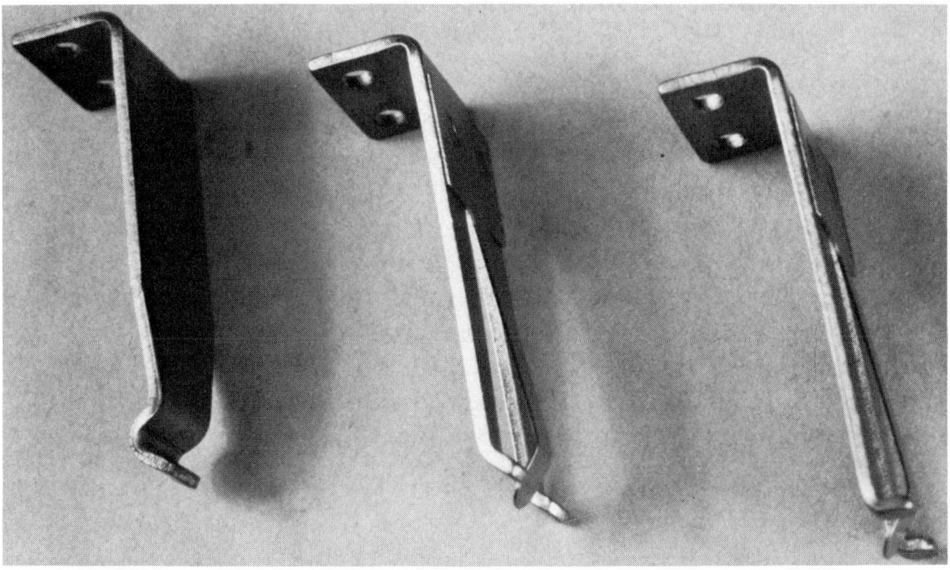

Figure 4.24 Three styles of conveyor fingers. (Courtesy of Hollis Automation Inc.)

warp and twist the rails. Any warpage or twisting is detrimental to the proper conveyance of material because it creates nonparallel conditions and the potential exists for pwbs to fall off the conveyor. Wear strips are also valuable to prevent conveyor wear from the finger pressure and should be considered in the design.

The mechanism for adjusting conveyor width should be protected in its entirety to maintain the tolerance accuracy necessary for proper width control. The location of the adjustment mechanism should be in a place not conducive to spillage from the process, e.g., in back of the fluxer between fluxer and preheaters, or in the back of the solder pot where solder, oil, or dross can drop off onto the mechanism. Also, if the solder equipment has either a foam or spray fluxer and uses oil in the solder pot, the spray and oil overflow will create a mess over the entire assembly, requiring extra maintenance and care in operation of the system.

4.5.3 Conveyor Drive Mechanism

Drive mechanisms must have variable-speed adjustments capable of speeds up to 16–20 ft/min. The drive must be torque-controlled to maintain consistent speed as the conveyor is loaded with product. Therefore, this will keep the drive system accurate over the entire span of the operation, from an empty conveyor to a fully loaded conveyor.

Traditional drive mechanisms have D.C. drive motors which lose speed as the conveyor is loaded with product. This conveyor speed slowdown introduces another variable which is detrimental to the soldering results. It affects the time in the solder, and with a slower conveyor speed, the incremental increase could be in the range of 1-2 sec. The impact could potentially add pwb delamination conditions to soldering types of problems, i.e., icicles, dewetting, nonwetting, etc.

4.5.4 Fixture Conveyor

Fixture conveyor, as its name implies, is designed to transport fixtures across the automatic soldering system as displayed in Fig. 4.25. The pwas are mounted into the fixtures, which are specially designed to encompass an array of variables yet are fabricated with material durable enough to withstand the environment of the solder process.

Fixture conveyors can be attached directly to the system or can stand alone in respect to the soldering equipment. Particular attention must be paid to standardizing the fixture height from the top of the conveyor rails. This is a must for all fixture type of conveyors.

On fixture conveyors, the fixtures are usually attached to a chain on the top of the conveyor rail. The dimension from the top of the rail to the top of the solder pot must be kept constant to certify correct contact areas with the solder pot. Also, a consideration with the fixture conveyor is the smooth sliding of the chain. It must be properly lubricated to certify that it doesn't have any jerking motion which could affect the soldering results and the point of solder contact.

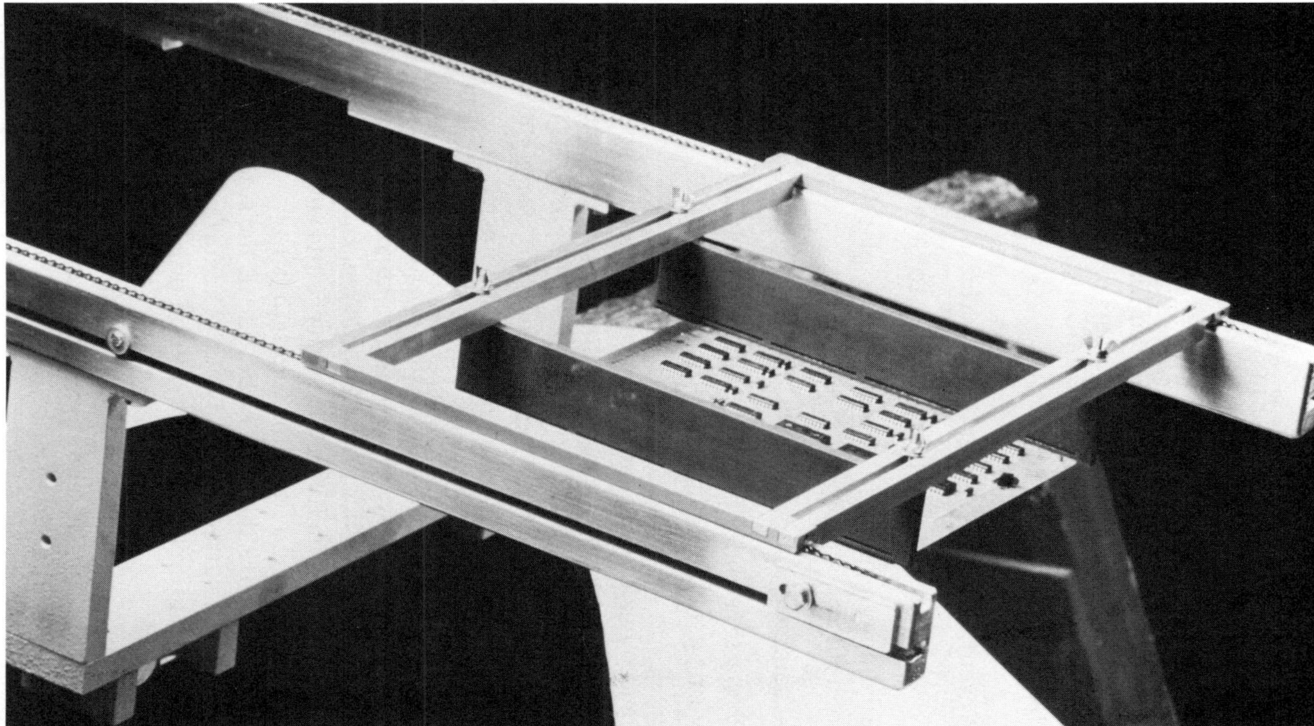

Figure 4.25 Fixture mount on a fixture conveyor. (Courtesy of Hollis Automation Inc.)

4.6 FIXTURES

By definition, fixtures are used as a means of holding an object for the purpose of performing a function on that object. This simple statement has led to sophisticated designs in fixturing. Together with soldering systems, fixtures have always played an important part in the following areas:

- Preventing excessive warpage of printed wiring boards
- Preventing gold fingers from being soldered
- Holding nonstandard components in place during the soldering operation
- Accommodating various outline configurations
- Preventing solder from coming over the top side of printed wiring assembly

Fixturing also has the dubious honor of being either specifically designed for unique application or being universal in nature. Unique fixturing is expensive to fabricate and purchase in addition to the total number that may be necessary for high-volume production. Universal fixturing must be readjusted for all assembly styles which will be soldered. This is time consuming and more deleterious to the life expectancy of the fixtures.

Soldering product with fixtures also creates a large heat-sinking effect in the solder pot. This can cause temperature reductions and affect the quality of the solder joints and increase touch-up. One method to monitor this function is to have two thermocouples mounted into the solder pot, one in the sump itself and the second on the surface of the wave. The difference between them can be monitored and reacted upon quickly when major variations occur.

Consistency of fixture materials must be considered when a variety of fixture suppliers are used. This consistency must not only include the materials used but also the dimensional accuracy of all the critical dimensions.

Materials for fixtures must be coordinated very closely with the flux chemicals being used. Aluminum plating, anodized or hard-coated, stainless steel, and cold-rolled steel are all potential candidates for fixtures. However, with the proper flux these materials can also be soldered, a condition which is unacceptable for any fixture. Some water-soluble fluxes will solder stainless steel and must be taken into consideration when designing a fixture.

Dimensional accuracy relates to the consistency of the location of the assembly in the conveyor finger or fixture as compared to the surface of the wave itself. Any variation will create different contact forces which will impact soldering quality results. The ability of the fixtures to remain within dimensional tolerances during the soldering cycle is important. Soldering temperatures and related materials are dangerously high and corrosive, and the environment to which they will be subjected must be understood when designing the fixtures. Multipurpose fixtures are used to protect certain areas of the module from soldering, e.g., to keep the module from warping and twisting and allowing for the expansion of the module during the solder cycle. During the cleaning stages these fixtures must not entrap or prevent the soldering chemicals from being cleaned (see Fig. 4.26).

With all these issues, fixtures must be easy to use and store as well as impervious to the handling and abuse which they will encounter during usage. The material coating or anodizing of aluminum fixtures must be designed to withstand the cleaning process. The alkalinity of the detergent systems will remove and destroy the coating and some fluorinated solvents, if not properly controlled, will attack the aluminum itself.

Once the phenomenon starts, as the fixtures pass over the solder pot, a certain percentage of the fixture will dissolve into the solder pot, creating a rise in the elemental contamination in the solder pot.

In summary, fixtures must be designed taking into consideration the following:

Figure 4.26 Fixture showing the adjustment bars for the various width boards. (Courtesy of Hollis Automation Inc.)

MECHANICALLY

- Height of the wave in relationship to the conveyor
- Ability to absorb the expansion of the module without allowing it to warp and twist
- Prevention of certain sections being soldered
- Easy adjustability for various configurations
- Allowance for quick and accurate placement of module into the fixtures and allowance for quick and easy removal
- Thermal stability
- Weight of the product
- Heat sinking of the solder pot

CHEMICALLY

- Compatibility of fixture finish to fluxes, solvents, detergents, and tinning oils

The author prefers black anodized aluminum plate for dedicated fixturing and black anodized aluminum bars for universal fixturing. These materials have proven themselves over the years. As for the materials that do not contact the wave or the fluxer, but do contact the cleaning cycle, stainless steel works very well. When designing a fixture, keep in mind the closeness of the components to the edge of the fixture. If the components are too close to the edge, the potential of missing solder joints or empty-plated through holes is very high. As shown in cross-section in Fig. 4.27, the solder will not wet up into the corner. The fixture must be relieved to allow the solder to create the joint when the component is near the edge of the board. This can be seen in Fig. 4.27. This relief allows the solder to contact the component and create the joint. A point to consider, however, is that the metallic mass of the fixture will draw heat away from the board during the preheat cycle and solder operation, which could cause poor

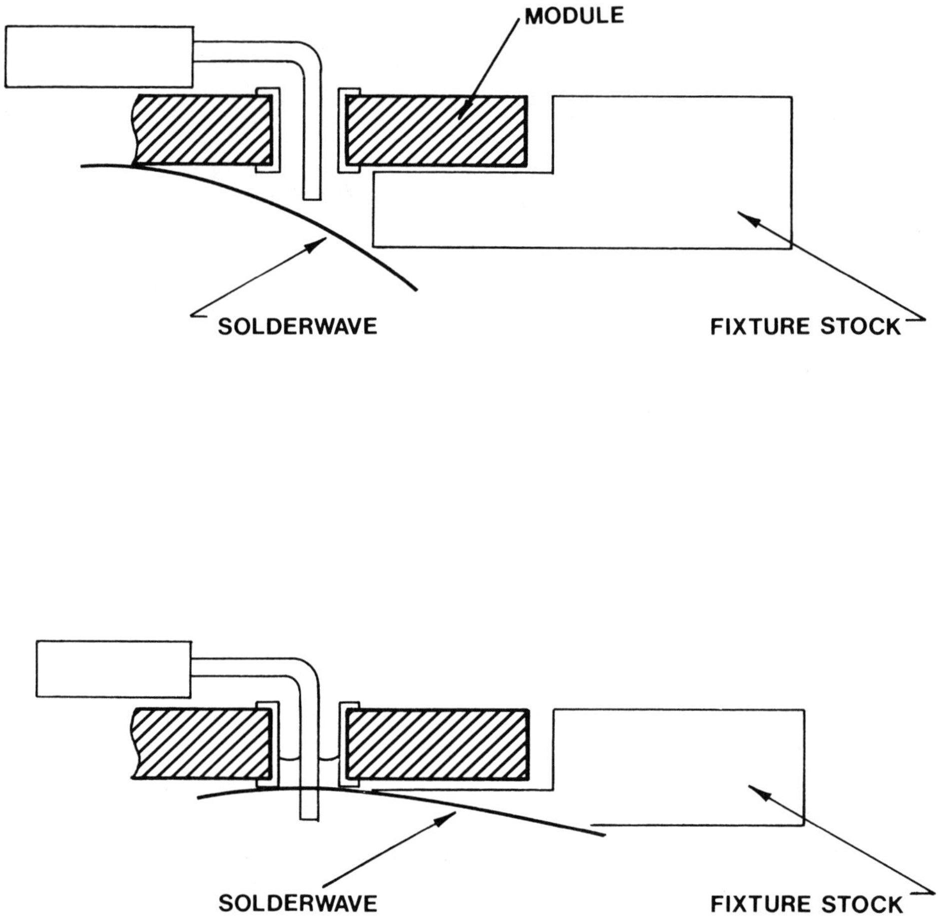

Figure 4.27 Module assembly sitting in a fabricated fixture and modified for ease of soldering.

solder joints. This must be taken into consideration when developing the parameters of the soldering process.

4.7 EQUIPMENT JUSTIFICATION PROCEDURES

Justifying equipment can be very involved. What I attempt to do in this section is describe one method that can be used to create an economic justification for an automatic wave-soldering system.

Manual soldering methods required approximately 2-3 sec/C joint, with an average of 200 joints per assembly on low-technology types, or 2000 joints per assembly on high-technology types. Hence, the total soldering time could have a range of 6–10 min for low-technology and 60–100 min for the high-technology assemblies. Automatic mass soldering equipment relates the total soldering time to the conveyor speed; hence, a pwb 5 in. long, soldered at a conveyor speed of five per min takes 5 sec to solder, or a minimum improvement of 98% over manual soldering methods.

An example of developing the standards for mass soldering and comparing the results with manual soldering follows:

Board length = 12 in.

Board width = 10 in.

Average number of solder joints per in.^{2} = 20

Machine conveyor speed = 5 ft/min

Machine available width = 16 in.

Total board square area = 12 × 10 = 120 in.^{2}

Total solder joints = 120 × 20 = 2400

Machine speed: convert 5 ft/min to 60 in./min, to 1 in./sec.

Therefore, with the linear length of the pwa being 12 in., the total soldering time is 12 sec. Hence, in 12 sec., 2400 joints were created. Stated differently, this equates to 200 joints/sec.

The manufacturing cost advantages are evident immediately; however, there are other advantages, i.e., consistency of solder joints on each and every assembly, improved quality, and process controls.

4.8 PROCESS CAPACITY DEVELOPMENT

Process capacity can be determined using similar logic. Knowing pwa length and the spacing between each, an effective product length can be utilized for this model. For example,

Pwa length = 12 in.

Pwa spacing = 3 in.

Total effective board length = 15 in.

Conveyor speed = 5 ft/min or 60 in./min

Define the capacity or utilization of the machine.

60 in/min ÷ 15 in/board = 4 boards/min

4 boards/min (X) 60 min/hr = 240 boards/hr

240 boards/hr (X) 8 hr/day (X) 5 days/week (X) 52 weeks/year = 499,200 pwas/year, based on 100% efficiency

In conclusion, any efficiency modeling that may be used is then based on 499,200 boards/year. So, if the schedules called for 50,000 pwas/year, the efficiency of the system would be a ratio of soldering 50,000 units, to an established capacity of 499,200. In this example, the utilization of the system is 10%, and the equipment would be idle 90% of the time. This is poor utilization of assets.

ASSOCIATED READINGS

1. Jim Moore, *A Solder Reflow Review*, Insulation/Circuits, March 1978, pp 49-52.
2. H. Terrance O'Rourke, *Solder Coating and Hot Air Leveling*, Hollis Engineering Co., Nashua, NH.
3. James Langan, *How to Achieve Quality Tin-Lead Plating*, Insulation/Circuits, March 1979, pp 31-35.
4. *Which PC Coating Stay Solderable Longest*, Circuits Manufacturing, September 1972, pp 18-27.
5. D. Schoenthaler, *Solder Coating Thickness Considerations for Hot Gas Solder Leveling*, Insulation/Circuits, November 1978, pp 39-43.
6. Hani Tawil, *Coat, Plate or Plain? A study of Surface vs. Solderability*, Insulation/Circuits, June 1976, pp 27-29.
7. Nicholas J. Spilliotis, *Tin-Lead Plating Practices that Assure Proper Wetting Upon Solder Reflow*, Insulation/Circuits, June 1978, pp 33-36.
8. Norbert Rehbach, *Hot Tinning of Printed Wiring Boards*, Electri-Onics, February 1983.
9. A. Rahn, Ph.D., *Hot Air Leveling Applications*, Circuits Manufacturing, March 1981, pp 20-24.
10. *PC Solderability Design and Pretreatment Prove Critical*, European Institute of Printed Circuits Seminar on Solderability, Circuits Manufacturing, January 1979, pp 10-20.
11. *Tin and Tin Alloy Coating*, IPC Joining Handbook 2.4.2.2, March 1975, pp 1-10.
12. C. J. Thwaites, *Soft Solders and Soldering in the Electrical Industry*, Tin Research Institute.
13. Robert H. Gengler, *High Quality Mass Soldering for High Density PCB's*, Insulation/Circuits, October 1981, pp 50-54.
14. Matthias F. Comerford, *The Debridging of High Density PWBA's by the GBS System*, Nepcon West Proceedings, 1982.
15. M. F. Comerford, *GBS A New Dimension in Wave Soldering*, Proceeding of PC 79.

16. T. Y. Chu, *A General Review of Mass Soldering Methods*, Insulation/ Circuits, November 1976, pp 73-75.

17. Joe Keller, *Soldering Without Shorts, Joints Are Auto ND Tested*, 1981 Proceedings Annual Reliability and Maintainability Symposium, pp 1-16.

18. Howard H. Manko, *Understanding the Solder Wave and Its Effects on Solder Joints*, Insulation/Circuits, January 1978.

19. Paul A. Thibodeau, *Computerizing the Wave Soldering Process*, Electri-Onics, April 1985, pp 37-40.

20. Edward F. Duffeck, *Soldering Process vs. The PC Board*, PC Fab, October 1983, pp 67-70.

21. Leo Lambert, *Air Knives Have an Edge on Solder Defect Control*, Electronic Packaging and Production, February 1984.

22. Gary Barrett, *The Effect of Adding a Post Soldering Hot Air Knife*, Circuits Manufacturing, January 1983, pp 48-52.

23. Tom Dixon, *Automatic Controls Step into Wave Soldering*, Electronic Packaging and Production, August 1983, pp 52-55.

24. Dallas Erickson, *Trends in Automatic P.C. Soldering*, Electronic Packaging and Production, June 1981, pp 65-86.

25. Dr. Ira B. Goldman, *Differential Thermal Analysis as an Aid to Solder Flux Characterization*, Insulation/Circuits, March 1981, pp 41-44.

26. Dr. Elias Kimmel, *Fusible Temperature-Indicating Materials*, Electronic Packaging and Production, October 1980, pp 173-178.

27. D. R. Wallis, *Trends in Testing for Wettability*, Brazing and Soldering No. 3, Autumn 1982.

28. John C. Mather, *Need for Tighter Requirements on Circuit Board Solderability*, IPC Proceedings, September 1983.

29. John C. Mather, *Circuit Board Solderability*, PC Fab, April 1984, pp 34-49.

30. Helmar Ulbricht, *Surface Resistivity in Printed Circuits*, Metal Finishing, September 1983, pp 43-47.

31. Gert Becker, *Scanning the Solderability of a Surface*, Welding Research Supplement, October 1981, pp 202-206.

32. Ralph Woodgate, *PWB Solderability Does Anybody Care?* PC Fab, August 1984, pp 60-61.

33. Dr. Wallace Rubin, *Soldering Techniques Adapt to Advancing Technology*, Electronic Packaging and Production, December 1983, pp 80-82.

34. B. M. Allen, *The Kinetics of the Soldering Process*, Electronic Packaging and Production, July 1981, pp 288-293.

35. Duncan Mackay, *The Meniscograph Method of Solderability Measurement*, reprinted by Hollis Engineering Inc., date unknown.

5

Cleaning Flux Residues

5.0 CLEANING FLUX RESIDUES

Throughout the electronic manufacturing industry the main questions being asked are, "How clean is clean" and "What solvent should be used to clean the assemblies?" These questions have been the impetus of many experiments and studies related to electronic manufacturing. Experiments by chemical manufacturers and environmentalists have examined and argued to the point that much information is known about cleaning printed wiring assemblies. Few, however, wish to risk the impact of not cleaning.

Should we clean? And if so, which solvent is best to remove flux residues? Each solvent has certain unique disposal requirements which must be understood in detail to help forestall potential class action suits based on improper disposal or personal injury. Not removing the flux residues may cause electrical malfunctions that may be difficult to identify and repair, which will increase the cost of the product and damage customer relations. These issues must be reviewed and understood prior to developing any cleaning program.

Certain fluxes activators if left on the surfaces of printed wiring assemblies will create potential electrical failures. Since most activators are inorganic, they will ionize in the presence of moisture. They will also chemically change from being soluble in water to metallic salts. Some of these are insoluble in typical flux removal chemicals. Although insoluble, however, these materials will activate when combined with moisture in the presence of an electrical potential between adjacent etch lines or metallic conductors. Activation creates dendritic growth between the metallic surfaces, which in turn causes short circuits. The dendrites have been shown to grow between etch lines within 3 min in severely corrosive environments. The offending activators must either be removed entirely from the surfaces of the boards or omitted from the basic flux material.

5.1 WHY CLEAN?

Cleaning has been performed for many reasons, some of which make sense and others which are suspect. However, through electrical improvements in circuitry and performance of systems, the cleaning process must be critically reviewed before implementation. Improper flux removal processes will impact electrical product performance and reliability.

There was a time when cleaning was performed primarily to ease handling or prevent electrical tester problems. Today, considering the effects on insulation resistance, ionic contamination and laminate degradation are factors if the products are left uncleaned. The impact of cleaning, whether to clean or not to clean, and to determine how clean is clean, is left to the discretion of manufacturers. They alone should know their individual product requirements, and as the theme throughout this book indicates, the knowledge of the product is a necessity to produce it in quantity with quality.

Much has been published covering the cleaning of printed wiring boards. There is general agreement that any contaminants that touch the boards must be evaluated for their potential impact on the reliability of the product. If any are found that will cause reliability or functional problems, they must be removed.

Before developing a cleaning process and determining the materials and equipment to use, the contaminants must be identified. This can be accomplished by performing laboratory analysis on the residual contaminants or questioning the vendors about the residual materials after the soldering process. These materials may change chemically due to their exposure to the soldering temperatures. Once these residues have been identified, then the proper material can be chosen for the cleaning process.

5.2 DIFFERENT CATEGORIES OF CONTAMINANTS

The different kinds of contaminants are described in the following paragraphs.

5.2.1 Particulate Matter

This material is deposited as discrete separate particles which generally do not adhere to the board or assembly surface or to each other. Particulate matter includes dust, metallic chips, and other solid particles. The particles themselves may be conductive, e.g., copper slivers, or nonconductive, e.g., epoxy board chips.

5.2.2 Films and Residues

When liquids or pastes, such as oil, grease, fingerprints, chemical solutions, etc., are deposited as contaminants, they usually form a continuous wet or moist film. When allowed to dry, most films leave a residue. Generally these films or their residues are soluble; however, occasionally a film undergoes a chemical change and becomes insoluble (e.g., wax from capacitors, oil stains, etc.).

5.2.3 Polar Materials (Inorganic)

These include mostly salts, acids, and bases, each of organic or inorganic origin. The inorganic polar materials ionize in solution and become electrically active. It is this electrical conductivity of ionizable deposit that is detrimental to the field performance of boards. In the presence of humidity, conductive films are formed which provide paths for disruptive leakage currents.

5.2.4 Nonpolar Materials (Organic)

These materials may be solids or liquids and are mostly organic. Typical examples are oils, waxes, and rosin derivatives. These materials do not ionize and, if pure, may be classified as insulators. In practice, polar and nonpolar contaminants are generally interspersed or codeposited so that removal of both is required.

These are the typical contaminants that have to be removed from printed wiring boards that are processed through a manufacturing assembly line.

5.3 CLEANING MATERIALS

Two schools of thought exist about which material for cleaning is best: hydrocarbon solvents or water. Depending on which flux system is being used, be it either rosin or water-soluble, the flux removal materials and methods can be designed to remove the residues. However, before choosing the cleaning materials, identifying the contaminants and residues to be removed is a must. Consideration must also be given to the product design including areas of entrapment, compatibility of the components to the solvent being considered, and reliability requirements of the product. Another issue to consider is the product life and design, which may change and render the equipment ineffective. When this occurs, old equipment must be replaced or updated. As can be seen, equipment and material selection must be carefully evaluated to produce the desired results.

Cleaning equipment can vary from small, inexpensive batch-cleaning tubs to large expensive in-line conveyorized machines. Each system must be evaluated from the perspective of cost analysis and return on investment for the capital expenditure.

5.3.1 Cleaning Materials

Solvents, whether polar or bipolar, clean off the residual flux residues by dissolution. However, dissolution is a chemical cleaning action and will not remove anything that it does not dissolve. To remove insoluble soils, mechanical actions are necessary to dislodge and remove the contaminants.

Fluorinated Solvents

The major fluorinated solvents are divided into two groups: blends and azeotropes. Blends are typically bipolar solvents and azeotropes are polar solvents. Bipolar solvents will dissolve both polar and nonpolar contaminants, while polar solvents will only dissolve polar contaminants. Blends were designed to remove both the flux material and its activators. Flux is a polar material and the activators are nonpolar; therefore a blend offers a method and material to remove the residues.

Blends are a combination of two solvents which have different boiling temperatures. Precautions have to be taken when using blends because of the dual boiling points. Blended solvents will eventually separate during the boiling cycle in degreasers and reduce the dissolving power of the solvent to a level where the product will not be clean. In solvent blends, the two materials have different boiling temperatures due to their latent heat of vaporization. When exposed to elevated temperatures the two materials start vaporizing at their respective boiling temperatures. When this occurs the lower boiling solvent will evaporate faster, changing the overall mixture of the blend. At this point the base solvent reverts back to the original state, which may not be effective enough to perform the cleaning of the flux residue.

Azeotropic solvents have one boiling temperature and are easier to control than the blend over a period of time. In either case, blends or zeotropes, without stabilizers the solvents will attack aluminum and other sensitive metals. The typical stabilizer used is nitromethane and the product code labels are usually identified with the suffix S, i.e., Freon TMS, TES.*

Chlorinated Solvents

Chlorinated solvents are polar materials. The two most used are trichloroethylene and trichloroethane 111. Chlorinated solvents are very active and, if compared to the fluorinated solvents in the kauri-butanol values, they are more aggressive by orders of magnitude. Being more aggressive, they must be handled with more caution. Kauri-butanol is a type of grease and is used as a constant in measuring the cleaning effectivity of various solvents. This comparison is identified by the kauri-butanol number.

Detergent

The term *detergent* is a misnomer. Instead it should be *industrial saponifier.* The reasoning is that soap used in the home is called detergent and this material, although it performs the same type of function, is more alkaline in nature and more concentrated and therefore hazardous for home-type application. The saponifier is blended with water to create a highly effective alkaline cleaner. The bases of the majority of these saponifiers are amine hydrochlorides and surfactants. Not all vendors include foam suppressors as an active ingredient in the chemistry of the material. When the foam suppressors are not integral to the material, they must be separately added during use in the process. Saponification cleaning systems are designed strictly for removing rosin-based flux residues.

The system converts rosin to a rosin soap by the process of saponification with the "detergent." Saponification is the process of chemically converting a non-water-soluble rosin material into a

*TMS, TES, E. I. duPont trade name.

soluble rosin soap. The exact chemical conversion can be researched through the suppliers of the materials. Once the saponification has taken place the material is disposable in the municipal waste treatment system.

Although an effective material to use, its pH is extremely high, 12 or higher, even at low concentrations, and certain precautions must be taken for proper usage and disposal. Disposal is relatively simple. All of many "detergents" tried by the author were highly biodegradable, i.e., they broke down extremely quickly in the environment. However, due to the high pH of the effluent, it must be adjusted to the disposal requirements of the community.

Checking with local authorities regarding disposal should define the requirements that the community imposes on users of the local sanitary sewer system. An issue that is still being addressed for aqueous disposal is the quantity of heavy metals permitted in the effluent. The contaminants flux and oil from the wave solder machine are loaded with tin-lead, so high tin-lead levels can exist in the effluent. Effluents and wash tanks must be monitored regularly to determine the actual metal content. Although this number is typically within the guidelines established by the federal government, good neighbor policy dictates that communities be kept informed of industrial waste that will be introduced into their sewer systems. Issues such as board materials, components, equipment, location, environmental disposal, cleanliness requirements, product requirements, etc., must be addressed when boards are used and stored in a manufacturing environment.

As saponifiers are used in batch and in-line cleaning systems, the quality of the water should be monitored to assure that hardness is zero (0) grains per gallon. If the water in the community is termed hard (many grains per gallon of calcium carbonate), one may choose to soften the incoming water to zero grains per gallon by using industrial water softener systems available from companies like Bruner-Calgon or Culligan Inc. This will allow better mixing of the saponifier and water and provide cleaner boards.

Water

Plain water systems use either straight tap water or especially prepared water, such as deionized ("DI") water. The purpose of the water system is to dissolve all the contaminants on the surface of the boards and render them clean. Certain precautions must be taken, however, to increase the water's ability to penetrate small areas. High surface tension is inherent in any water system. At times special surfactants are utilized to reduce this surface tension, which allow for greater penetration and better removal of contaminants.

A product's design must take into consideration the cleaning process. Good design assures that parts and components are compatible with the process. Parts should be mounted up off the surface of the modules and typically any dimension greater than 0.007 in. is sufficient. This requirement is being reviewed carefully by the industry because of the use of surface mount components and their tolerances. Some of those devices are mounted flush to the board or epoxied to the board surface and cleaning beneath them is difficult if not almost impossible. A method being examined by industry is the use of nonactive fluxes in conjunction with higher solderability requirements, which will or should allow residues to be left behind, meaning not removed and not creating reliability problems. Water should be used only when it can be assured that no flux will be left on the board to cause corrosion problems and insulation resistance failures.

In addition, certain issues must be addressed so that the correct solvent will be chosen: either a hydrocarbon solvent or a water-type system.

5.3.2 Developing Data for Solvent Selection

There are many opinions regarding which solvent works best: chlorinated or fluorinated hydrocarbons, chlorinated, detergent, or plain water. Questions such as the following should be asked and answered to properly prepare the decision process:

1. Identify the contaminants to be removed. This will identify th solvent to use.

2. What is the base material of the printed wiring board? (Fr-4, phenolic, quartz polyimide, G-10.) This will identify some of the compatibility problems and porosity problems with the laminate materials.

3. Is there a solder mask on the board? Dry Film? Wet Film? UV Film? Incompatible cleaning material and uncured solder mask will exhibit some contamination problems and visual degradation.

4. Is the solder mask on side 1 (component side), side 2 (solder side), or both? This is critical to determine laminate and solder mask problems. Additionally, the reaction of solvents to the exposed etch may cause visual characteristics which may not be acceptable.

5. Does the board have plated through holes or is it single-sided? Single-sided boards have exposed laminate fibers within the drilled holes and are potential areas for entrapment of flux and cleaning chemicals.

6. Are components mounted flush to the board? This is necessary to determine the entrapment of contaminants.

7. Is the density of components to board surface area high? This will impact the ability of the solvent to be directed into the tight areas and result in some potential cleaning problems.

8. Are the components on one side or two? This is a concern with surface mount, because of the potential entrapment of chemicals.

9. Is there any oil used in the wave solder system? Is the oil compatible with the cleaning process? Is it synthetic and soluble in water or is it petroleum-based?

10. What is the time duration between the soldering operation and the cleaning cycle? This allows the flux to dry on the product and create difficulties in removal. If a water-soluble flux is used, the drying process converts the material from a soluble form to an insoluble form and creates a metallic salt which will absorb moisture from the environment and create electrical malfunctions.

11. Will this solvent be used for cleaning any other product that has not been wave-soldered? If yes, then the material must also be usable as a cold cleaner.

12. Should wave solder fluxes be compatible to the cored solder used in the manual soldering operations? It is not necessary to use two flux removal processes; however, if the hand-soldered product is cleaned in the same equipment, then the fluxes must be compatible.

13. With what kind of solvents are the components compatible? Define the compatibility of the components to the chosen solvents. This is important for hermiticity testing and marking permanence testing.

14. How much volume is the system expected to clean? The capacity of the system to clean the volume of product being manufactured is important to product flow through the process for work in process (WIP) control.

15. What is the expected capacity requirements of the system? Will the capacity of the equipment create future problems if the product mixes or volumes change?

16. Are there any environmental problems in the local area that must be addressed? Can the solvents be disposed of easily? Is there an approved vendor who can handle the waste? Remember that solvents are a hazardous waste and must be handled as such. If water cleaning is used, have the local municipalities been contacted to identify their requirements for sewer disposal and material limits. Typical municipalities have heavy metals, biodegradability, pH, phosphate, and volume restrictions.

The answers to these questions will help to identify the type of cleaning system and materials to use.

5.4 CLEANING METHODS

Three methods exist for cleaning printed wiring boards:

1. Manual
2. Batch

3. In-line

5.4.1 Manual Cleaning

Manual cleaning is performed with cold solvent and brushing techniques. If not performed correctly, this method can cause many problems. These occur when insufficient solvent is used to remove contaminants, simply spreading the contaminant over a larger area of the board. Water is usually not used in manual cleaning because the typical flux materials to be removed are not water-compatible. Additionally, if the flux material was water-compatible the cleaning cycle would take longer because of the inability of the water to dry quickly at room temperature. Also if a detergent system were utilized, to be effective it would have to be heated to 160°F (71°C), which is not practical for manual cleaning operations.

"Brush-and-bottle" techniques not only spread contamination over the surface, but when the brush is returned to the bottle, contaminants are introduced to the solvent source. After repeated cleaning cycles, the contaminants within the bottle are more concentrated than the flux that is being removed.

The best way to manually remove flux residues is to utilize a constant solvent supply mechanism and to work the solvent to the edge of the board. Efforts should be made to prevent any unnecessary spreading of solvent over areas of the board. Solvents are available in pressurized containers that serves this need well.

A disadvantage of manual cleaning is what happens to the flux residues on side 1 (or the component side) of the board. Accessing some areas on the component side can be difficult at best. Additionally, there may be solvent entrapment areas or component incompatibility. If this is a problem, the only solution is to use a flux which will deposit contaminants which are inactive under the operating parameters of the product. In most cases, this is accomplished by using rosin or rosin mildly activated flux to get a good solder joint. Whichever flux type is used, verify the activity of the residues by conducting experiments under various environmental conditions.

5.4.2 Solvent Batch Cleaning

Batch cleaning can be performed with either hydrocarbon solvents, water detergent, or plain water. Hydrocarbon solvents utilize vapor degreasers. These systems vary in size and capacity and must be chosen according to the size of the product and the volume to be processed. The procedure is relatively simple: boil the solvent, control the vapors, and lower the product into the vapors (see Fig. 5.1). After a given period of time, remove the product, allow it to dry, and the process is complete.

Solvent Batch-Cleaning Cycle

Cleaning in a vapor degreaser occurs by a condensing effect on the product lowered into the vapors. As the boards are lowered into the vapor section below the cooling coils, the vapor immediately begins condensing on the cooler surfaces of the boards and components. During this time frame, the solvent is dissolving the rosin from the flux system. If enough solvent condenses on the surfaces, the solvent will run off the board surfaces carrying away the contaminants.

Once the boards reach equilibrium temperature within the vapor section, the condensing stops and the cleaning cycle stops. To restart the cycle, either the spray wand is used to spray the boards with cool, clean solvent allowing the condensation process to start again or the boards are lowered into the clean solvent section of the degreaser, then brought back into the vapor zone, for condensation cleaning. Depending on the levels of cleanliness required, a degreaser could have as many tanks or sections as necessary to achieve the required cleanliness levels. The condensed vapors would be brought to the last tank in the cycle and each preceding tank would have a lower level. This would create a cascading system beginning with the cleanest tank to the boiling sump.

Venting of Solvent Batch Degreasers

When using batch degreasers, certain elements must be monitored to prevent harm to the operators and using excessive solvent. This is accomplished by proper venting. If any fumes or odors can

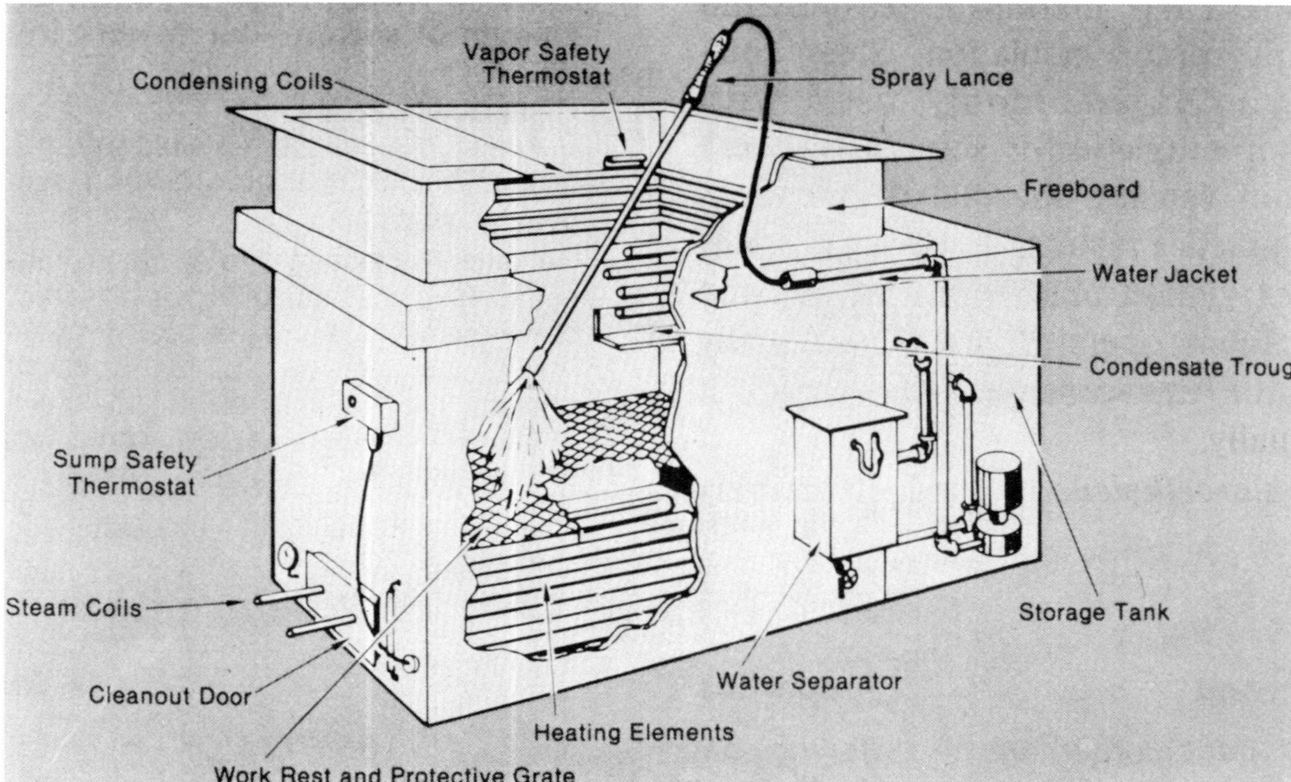

Figure 5.1 Sketch of open-top vapor degreaser. (Courtesy of Hollis Automation Inc.)

be detected around the system, something is wrong. The smell of vapors must not be detectable. Venting will remove the vapors if done correctly, implying that solvent vapors are heavier than air and an overhead venting system is useless and should not be constructed. Additionally, if an overhead venting system was to work efficiently, it would move the vapors and odor across the face of the operator, a dangerous practice which must be avoided. Venting must be done from the edges of the degreaser, pulling the fumes horizontally away from the top of the system. Once contained, they can be brought to any cleaning tower scrubber and vented outside of the facility.

Each solvent has a different toxicity level or TWA (time weighted averages), which is the maximum allowable daily exposure to the material vapors. There is a TWA number for each solvent type and this number must be monitored for proper operational safety controls (see Table 5.1).

Solvent Batch Degreaser Loading

Another consideration when using batch systems is the method used to lower the basket into the degreaser (see Fig. 5.2). If the rate of descent is too fast, it will push the solvent out of the tank, removing it entirely from the system. The entry rate must be as slow as possible to prevent this condition from occurring. Similar conditions apply to the removal rate. These can be controlled with mechanized cranes, programmed for each step in the operation, including dwell time in vapors and cool solvent tanks. With a crane system, the process is repeatable; and if properly established, quality of cleanliness will be consistent.

5.4.3 Aqueous Batch Cleaning System

Another type of batch system uses water. It can be used with or without detergent, and if used with detergent, it can be added automatically (see Fig. 5.3).

The basic aqueous batch system is nothing more than a standard dishwasher as shown in Fig. 5.4. The cycle steps are similar to the standard dishwasher including the drying; see Fig.

Table 5.1 Solvent Properties

Type of solvent	Toxicity (ppm) TWA	Flash point (TCC)	Boiling range (°F)	Freezing point (°F)	Water solubility % by weight at 68°F	Specific heat liquid at 68°F (Btu/lb/°F)	Latent heat at bp (Btu/lb)	Pounds per gallon at 68°F
Chlorinated Hydrocarbon								
Methylene Chloride	500	None	102-106	−142.1	0.2	0.28	141	11.07
Trichloroethylene	100	None	188-190	−123.5	< 0.1	0.23	103	12.22
1,1,1,-Trichloro-ethane	350	None	162-190	− 58	< 0.1	–	–	11.03
Fluorinated Solvents								
Trichloro-trifluoroethane (FC-113)	1000	None	117	− 31	< 0.1	0.22	63.12	13.06
Freon TMC[a]	600	None	97.2	−126	< 0.1	0.26	104.0	11.85
Freon TMS[a]	475	None	103.5	–	–	–	90.7	12.33
Alcohols								
Methanol	200	54	147-151	−144	–	0.60	473	6.61
Isopropanol	400	56	179-181	−126	–	0.60	287	6.55
Ethanol, 95%	1000	57	166-175	−198.4	–	0.62	–	6.76
n-Propanol	200	59	207-208	−196.6	–	0.57	296	6.74
Ketones								
Acetone	1000	0	132-134	−138.6	–	0.51	224	6.58

[a]Freon is Du Pont's registered trademark

Source: Adapted from EPA-450/2-77-022, November 1977, Control of Volatile Emissions from Solvent Metal Clean-and, and Du Pont's Technical Data Freon Solvent Properties, Bulletin No. FS-6.

Figure 5.2 Basket being lowered into degreaser. (Courtesy of Hollis Automation Inc.)

Figure 5.3 ECD aqueous cleaning systems. (Courtesy of ECD Inc.)

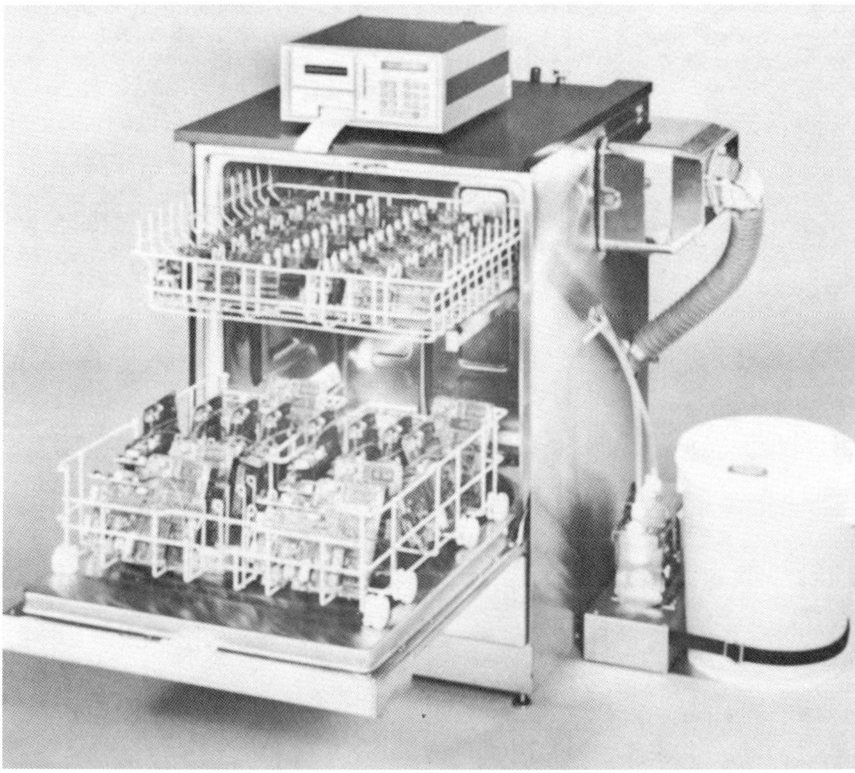

Figure 5.4 ECD aqueous batch (open). (Courtesy of ECD Inc.)

5.5 for a typical printout. On specialized batch units, the various steps are monitored and timed to accomplish their respective functions: prewash, wash, rinse, wash, rinse, rinse, dry. The prewash is used only when water-soluble flux systems are used in the soldering operation. When rosin flux systems are used, saponifier is automatically added to the first cycle and any other cycle desired (see Fig. 5.6). Printouts as shown in Fig. 5.7 display actual operations of the system. Rinse sections are most important in the detergent aqueous system. The removal of wash contaminants is a must to prevent redeposition of contaminants on subse-

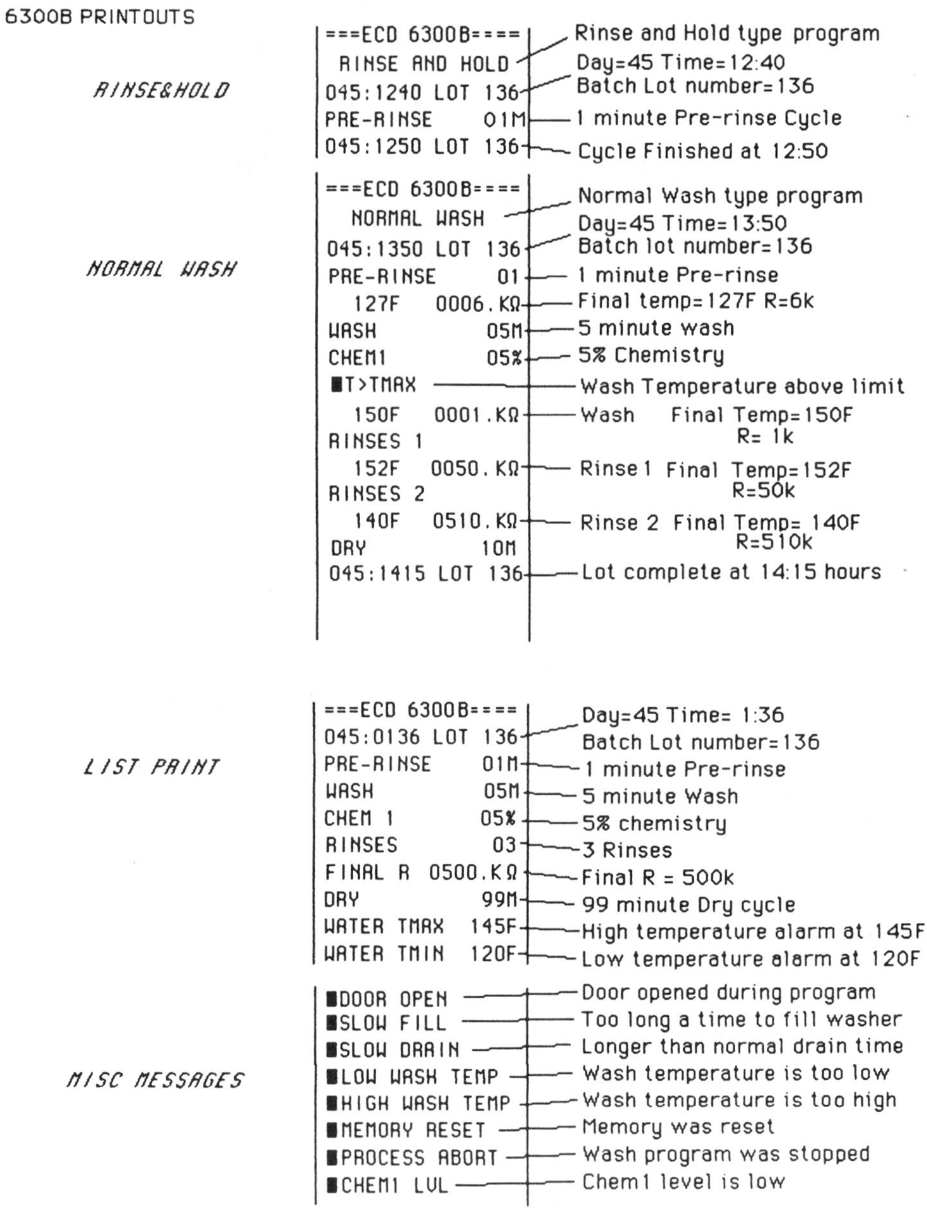

Figure 5.5 Computerized printout of cleaning cycle. (Courtesy of ECD Inc.)

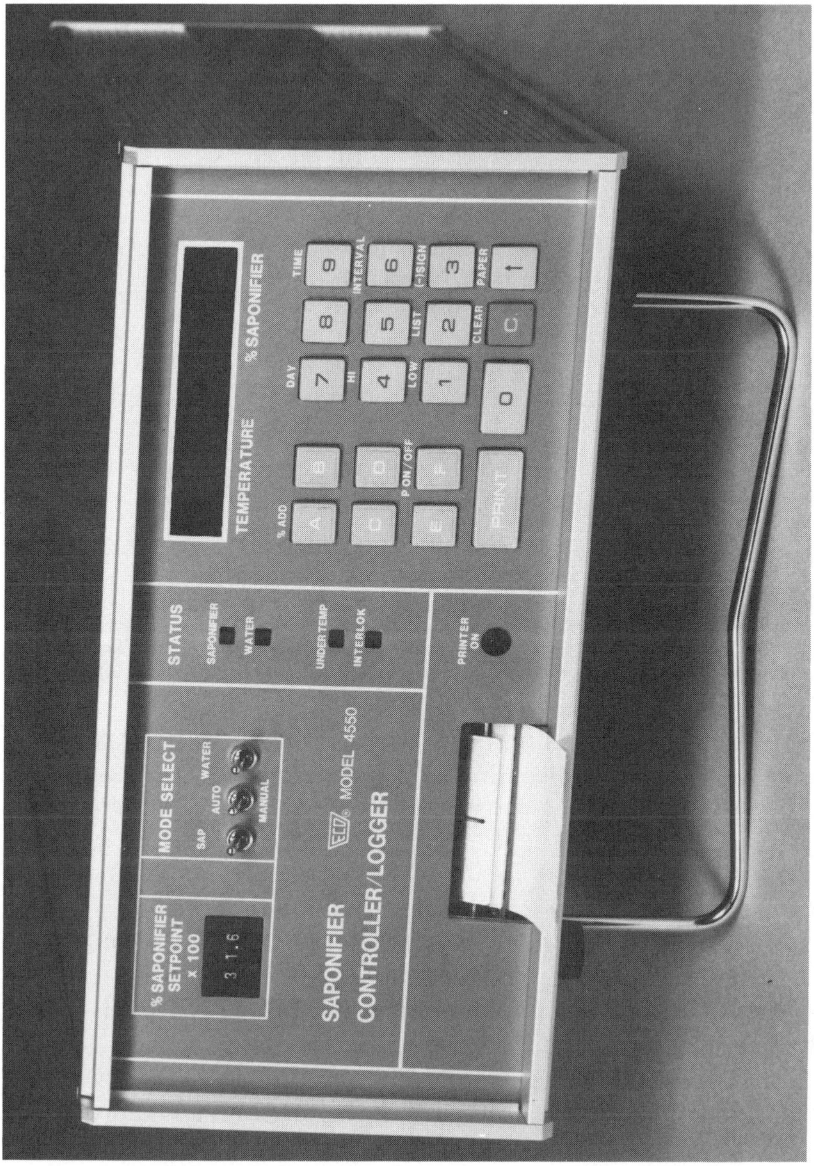

Figure 5.6 Saponified controller. (Courtesy of ECD Inc.)

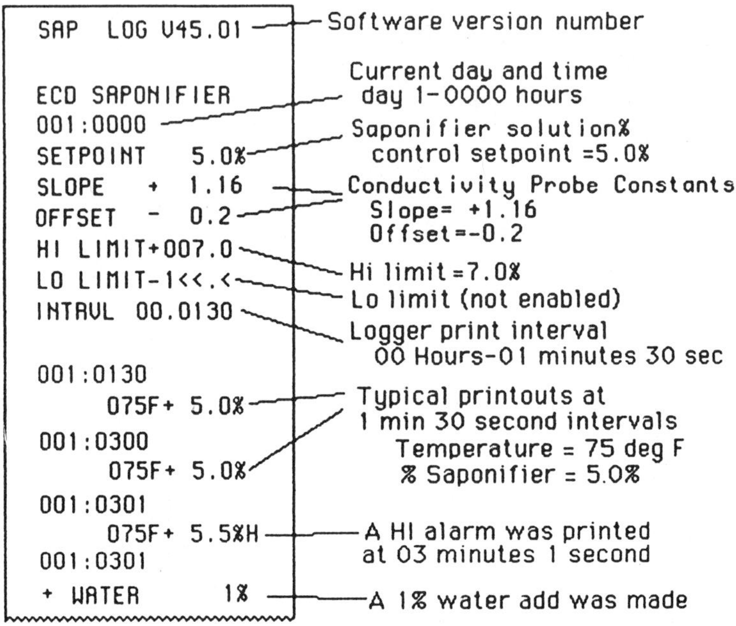

Figure 5.7 Control charts saponifier. (Courtesy of ECD Inc.)

quent wash cycles. The number of wash/rinse cycles is totally dependent on product cleanliness requirements.

Water Supply

Are there any fears of using plain tap water? Yes! Most municipalities are incorporating fluoridination programs by adding fluorides to the water systems. Fluoride is a halogen and reactive material.

The process of cleaning the base water supply is accomplished by adding chlorine. Chlorine is also a halogen and reactive. Therefore, when using tap water, have the water analyzed to exactly identify its contents so that the process can be properly defined.

Although these elements are present, do not despair; there are many companies that supply services to filter out the chemicals and many other type of minerals. As a point of interest, check with local municipalities on how many sources are available for local water, what the differences are between sources, and whether they switch from source to source without notifying customers. This will have a major impact on the cleaning process, if the water is switched from various sources without customers being made aware of the switch. Certify that you know the source of your water supply.

Quality of incoming water Quality of incoming water is a major factor for saponifier aqueous systems. Water hardness affects the final cleanliness of the product; therefore, it must be continuously monitored. Hard water, defined as grains per gallon of calcium carbonate, must be zero for proper intermizing of the saponifier in the water. Hard water prevents the detergent from going into solution, and therefore no saponification occurs between the rosin and the detergent. This causes a sludge to build up on all the surfaces exposed to the solution. This includes internal parts, such as pumps, nozzles, tubing, etc. This slude material can be removed by using an acid strip chemical such as Kenco Scale Strip (Kenco Industries, Inc., P. O. Bo 88500, Atlanta, GA 30338) or Lonco Mach 73 (London Chemical Co., 240 Foster Ave., Bensenville, IL 60101). This acid strip chemical removes the calcium sludge buildup in the aqueous cleaner system, including all the internal workings of the pumps, manifolds, and nozzles.

The use of these materials must be monitored closely due to the high levels of tin and lead within the sludge. Depending on frequency of use, the environmentally acceptable limits may be exceeded. If this occurs, special disposal techniques must be developed to prevent contamination of the environment. Test the solution for heavy metals, and if the metals are suspended solids, they can be separated through filtering or sedimentation. If they are identified as dissolved solids, a chelating agent can be used to create a precipitate. The sludge can then be handled according to local environmental requirements. Chelating agents are used to combine materials within the solution of saponifier waste. Knowing the contaminants in the aqueous solution, special chelating agents can be chosen and combined with the solution to precipitate out the contaminants.

Batch Aqueous Drying

The most problematic area of a batch aqueous system is drying. Although various attempts have been made to improve this cycle, much more work is still needed. The concept is to heat the environment and recirculate the air. This is all well and good; however, the humidity is so high within the chamber that an excessive amount of energy is required to dry the assemblies. The latent heat of vaporization of water, 972 BTU/#, means that 972 BUT's are required to vaporize 1 lb of water. *Note*: it takes 1 BTU to raise 1 lb of water 1°F, and therefore drying is very energy-intensive.

Batch drying also proliferates the need for clean water, with regard to mineral content. This has to be considered because in drying the module, the water will evaporate off the surface of the modules and the contaminants that are in the water will be deposited onto the surface during the evaporation cycle. These residues could affect the electrical parameters of the assembly. One way to prevent this is to use dionized (DI) water in all the rinse sections, especially in the final rinse cycle.

5.4.4 In-Line Systems

In-line systems can be used with either hydrocarbon solvents or aqueous chemistries. The advantage of the in-line system vs. the batch is throughput of product through manufacturing. Batch systems remain idle while the boards are being queued into lots, whereas in-line systems can process product without queuing, as shown in Fig. 5.8.

Solvent In-Line System

The generation and control of vapors in the solvent in-line cleaning system is basically the same as the batch machines. However, the mechanism of transporting the module through the system and the method of cleaning are different. In the in-line system the product is placed on a stainless steel mesh belt and transported through the system, as shown in Fig. 5.9. To prevent the fumes from escaping, the belt is inclined into the entrance and

Figure 5.8 Boards on conveyor belt. (Courtesy of Hollis Automation Inc.)

Figure 5.9 Vitronics aqueous cleaner. (Courtesy of Vitronics Inc.)

out of the exit section of the vapors. This incline is fixed and designed to prevent the boards from slipping on the belt while remaining in the solvent tank.

In in-line systems, as in the batch systems, the vapor area is also controlled with cooling coils and includes large recovery sections. Recovery is important to supply the spray nozzles that provide mechanical abrasion on the surfaces of the modules. This abrasion, in addition to the solvent dissolving power, enhances the cleaning process.

Stills

A still is usually purchased to augment the supply of fresh solvent to the degreaser. The still should be sized to generate enough fresh solvent that the spray nozzles can be used continuously. Stills are sized by the amount of gallons that they distill within a given time period.

Aqueous In-Line Cleaning System

Aqueous cleaners have two basic designs and the following generic characteristics:

- Detergent wash systems (Fig. 5.10)
- Water wash systems with prewash section (Fig. 5.11)

Detergent wash systems are designed to dissolve polar contaminants (rosin flux residues), whereas water wash systems are designed to remove strictly nonpolar (organic activators) contaminants.

The wash and rinse sections of all aqueous cleaners are designed to dissolve and remove nonpolar contaminants and mechanically remove non-water-soluble polar contaminants.

Aqueous Description—Operational Parameters

The major difference between each type of system is determined by the purpose of the various wash and rinse sections. Standard detergent systems have four wet sections (see Fig. 5.12) whereas water systems have five wet sections. Additional clean-

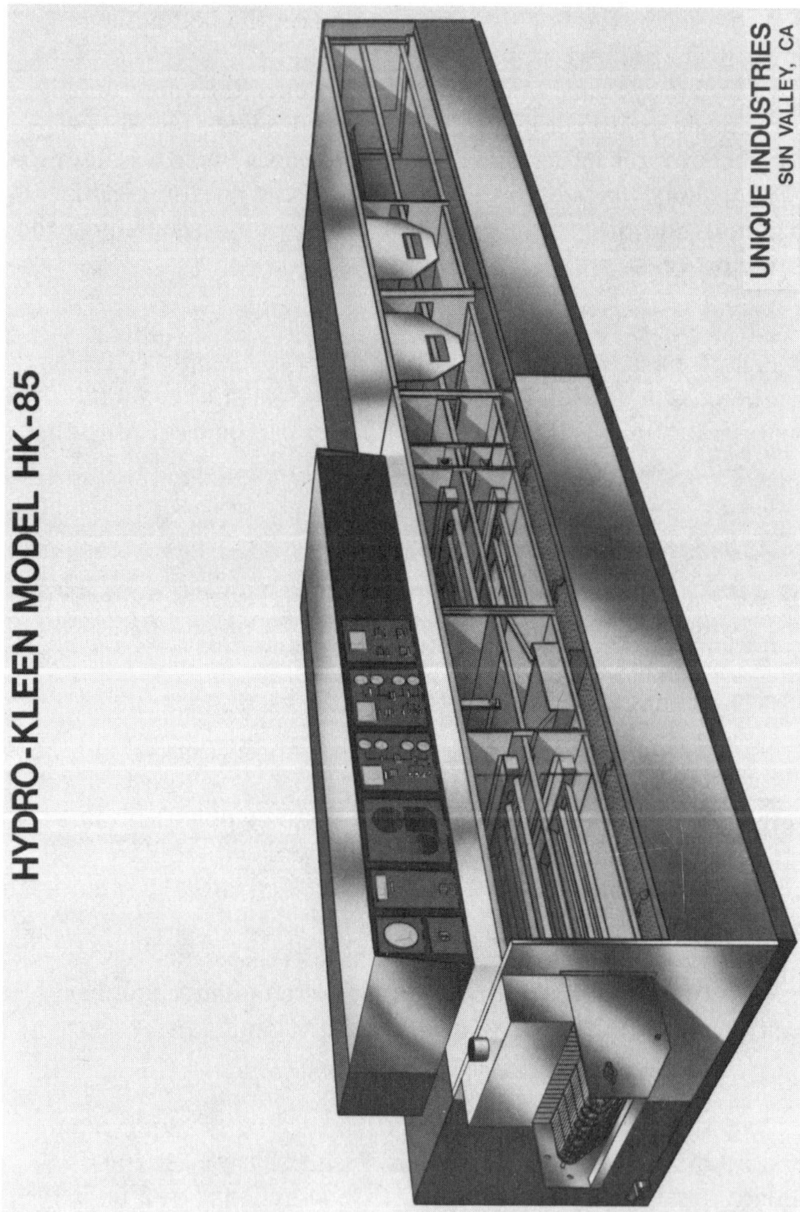

Figure 5.10 HK-85 without prewash. (Courtesy of Unique Industries Inc.)

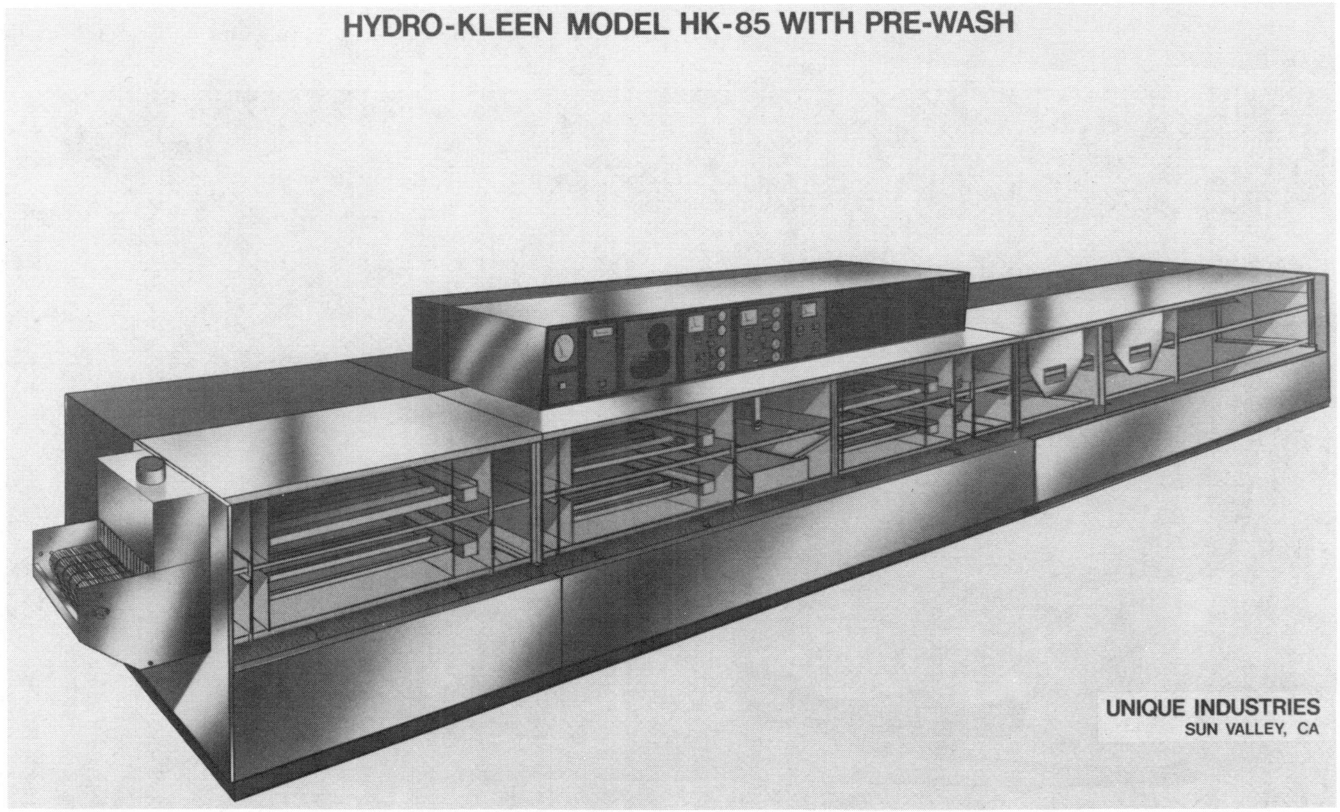

Figure 5.11 HK-85 with prewash. (Courtesy of Unique Industries Inc.)

Figure 5.12 Hollis polyclean. (Courtesy of Hollis Automation Inc.)

ing requirements can necessitate the purchase of extra optional sections. In detergent wash systems, the wash tank section is utilized to dissolve the polar (rosin) contaminants while the remaining sections perform rinsing functions.

In a water wash system, each section has the capability to dissolve and remove the nonpolar (activators) contaminants; however, it must be able to access restricted areas, such as the tight spacing beneath components. Under no conditions should it be assumed that all contaminants will be removed. The only contaminants that will be removed are those accessible to the solvent medium; in this case, water. The prewash section is a necessity in systems used for removing water-soluble flux residues. Many of the fluxes and wave solder oils are manufactured with surfactants to decrease surface tension increasing its ability to wet the parts being soldered. They also happen to be large foam generators, implying that when mixed with water, large quantities of foam develop. Therefore to prevent foaming within the system, a prewash section is used to flush away those contaminants prior to the product entering the recirculating wash section.

Also in the water wash system, a cascading concept is utilized to make use of the dirtiest water in the first rinse, then progressively cleaner water in the succeeding rinses. This allows less redeposition of contaminants from the sprays of each section. The final rinse must be the cleanest because if the water dries by evaporation, any material remaining in the water will be deposited onto the modules.

What Makes an Aqueous Cleaner?

This section will describe the subsections such as pumps, motors, nozzles, manifolds, pressure gauges, water level sensors, and other elements which constitute an aqueous cleaner.

Motors and pumps Motors and pumps, as shown in Fig. 5.13, are used to move the water through the manifolds or nozzles of the system.

- ○ Wash pump motor–1.10-1.5 hp
- ○ Rinse pumps motor–1.0-1.5 hp

Figure 5.13 Pumps and piping. (Courtesy of Unique Industries Inc.)

- Prewash (recirculated option) pump motor—1.0-1.5 hp.
- The pumps attached to the motors vary considerably from system to system. The best are stainless steel, which is impervious to cleaning detergents. Plastic base materials pumps are subject to being affected by the chemicals and destroy themselves through cavitation.
- Ventilation blowers are never supplied with the system. Usually, a 5.0- to 7.0-hp-high turbine blower supplied by the user is mounted on the roof of the building.
- Conveyor drive motor.

Existing manifold pressure control meters The piping scheme used for moving the water to various sections of the wash tank are called manifolds, as can be seen in Fig. 5.14. To dispense the water, either slots of predetermined width and depth, or nozzles sized to various pressure and volume are used in the manifolds.

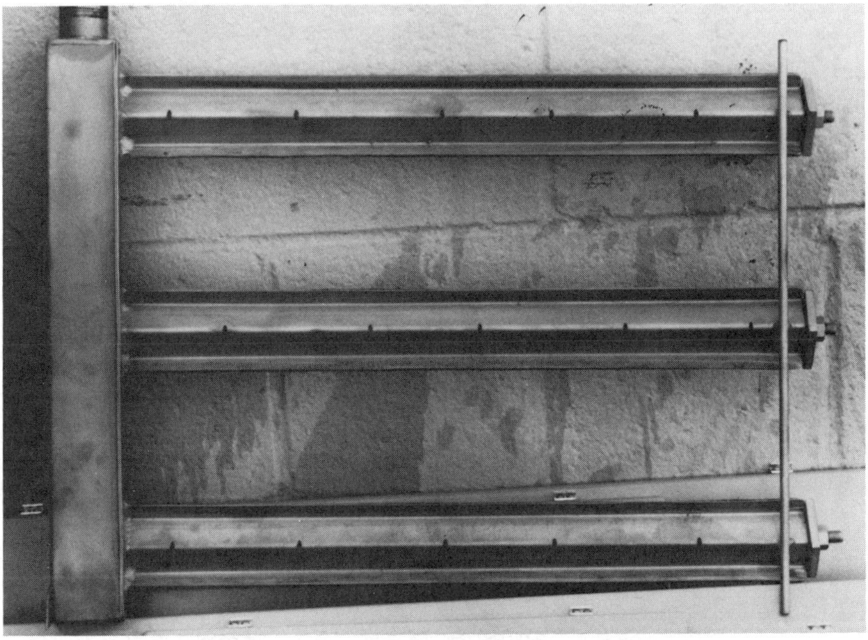

Figure 5.14 Manifold system. (Courtesy of Unique Industries Inc.)

These cutouts and nozzles are what identify the differences between high-volume/low-pressure and low-volume/high-pressure systems.

Other configurations utilized are standard-type pipes (see Fig. 5.15), be they either metal or plastic, fitted with various configurations of nozzles. Whichever method is utilized, the main criterion is total spray coverage of the belt and enough volume and/or pressure to dissolve the contaminants (see Fig. 5.16).

The meters and gauges identifying the manifold pressures are located in the following areas. Note that the meters and gauges are standard liquid pressure gauges (see Fig. 5.17).

Prewash section

○ upper

Figure 5.15 Pipe manifold system. (Courtesy of Hollis Automation Inc.)

Figure 5.16 Spray manifold system. (Courtesy of Hollis Automation Inc.)

Figure 5.17 Pressure gauge controls. (Courtesy of Unique Industries Inc.)

○ lower

Wash section

○ upper

○ lower

Drag-out (1st rinse)

○ upper

○ lower (optional)

Recirculated rinse (2nd rinse)

○ upper

○ lower

Final rinse (3rd rinse)

○ upper

○ lower

Water Level Sensors Each recirculating tank has water level sensors. They are all similar in style and perform the function in a similar fashion. Electrical probes of various lengths are mounted into the solution tanks where electrical contact is made through the dielectric of the tank solution between any two probes, and an action occurs.

In the wash tank, this action is threefold:

1. Low level––Certifies enough solution is in the tank to activate the heater coils.
2. Intermediate level–Certifies enough solution to activate wash pump. It is also used for determining the need for water makeup.
3. High level–Shuts off incoming water supply.

In the rinse tank, the action is twofold:

1. Low level–Certifies enough solution to activate the pump and motor. It is also used for determining the need for makeup water.
2. High level–Shuts off incoming (fill) water.

If the rinse tank is equipped with heater coils, then an additional sensor is utilized to prevent the heaters from being activated without any water in the tank.

Temperature controllers Standard equipment uses two controllers in the following sections.

- Wash tank
- Oven

Depending on the type of equipment, both controllers are used to initiate set points and monitor/control the set point. In other systems, the oven units only monitor oven temperature and are wired with a defined high-temperature limit to prevent system overheating.

Feed Mechanism Water inputs into all aqueous cleaners use standard plumbing. However, the quantity of incoming water lines may vary from system to system. Older systems have one incoming water line, feeding the supply to the wash and rinse tanks, and the final rinses. Depending on supply and volume, when wash and rinse tanks demand water, they impact the final rinse by lowering its pressure and volume. To this end, some systems are now equipped with dual incoming lines for initial supply and final rinse manifolds. This dual supply also lends itself to the use of DI water in the final rinse. In cascading systems, this works out perfectly because the DI water is only supplied throug the final rinse manifolds or nozzles while the initial supply in the final rinse tank is standard tap water. The water in the final rinse tank is the supply used for the preceding rinses and to use DI water for the total initial supply is costly and not effective. Also, the use of pretreated water, be it softened, heated, or both, is easily accomplished with the dual-input line system.

Feed Controllers These are typically electrical gate switches activated by the conditions of the water levels in the tanks. It is important to note and monitor the following:

1. Incoming water volume in gallonage, to determine the quantity of saponifier to be added for maintaining proper concentration
2. Water flow going into final rinse to monitor total water usage in the system

Drain Mechanism Drain systems are relatively simple. Gate valves in both wash and recirculated rinse tanks are manually opened to remove material from the tank.

Drain Control It would be beneficial to monitor the total effluent leaving the system and compare it with incoming volume. These two numbers must correlate; if they don't, then something is wrong and the system must be shut down. This follows the premise that what goes in must come out, and hopefully it can be sent out in a controlled and defined manner.

Detergent Concentration and Rinse Water Cleanliness pH sensors or conductivity meters are used to monitor and determine the conditions of the solution in the wash tank. They are not used for controlling the amount of solution in the tank. However, they are not functional!

The pH of detergent is in the 11-12 range and it is very concentrated. Minor amounts in the wash tank reflect the high pH numbers. This implies that although the concentration is down and the cleaning effectivity reduced, the pH still appears to be acceptable, when in fact the pH is irrelevant to the amount of detergent in the wash tank.

As for the conductivity meter, it is calibrated by adjusting it to a predetermined setting after the correct percentage of detergent has been added to the wash tank. The premise is that as the concentration is reduced, so will the conductivity be reduced. As the detergent concentration is depleted, however, it is replaced by the oil and flux from the soldering operation, which maintain the conductivity; hence, the conductivity remains the same, and the detergent concentration is depleted resulting in dirty boards, without impacting a change in the meter reading.

Rinse water cleanliness is measured with conductivity meters. This is only a monitoring function to identify the condition of the water in the recirculation rinse tank. The purpose for using the meter is that the boards are only as clean as the final rinse; hence, monitoring this section certifies good process control of the system.

Aqueous Description and Typical Operating Functions

The various sections of the aqueous cleaners can be seen in Fig. 5.18. They are as follows:

1. Entrance
2. Prewash (option)
3. Recirculated wash
4. Rinse to drain

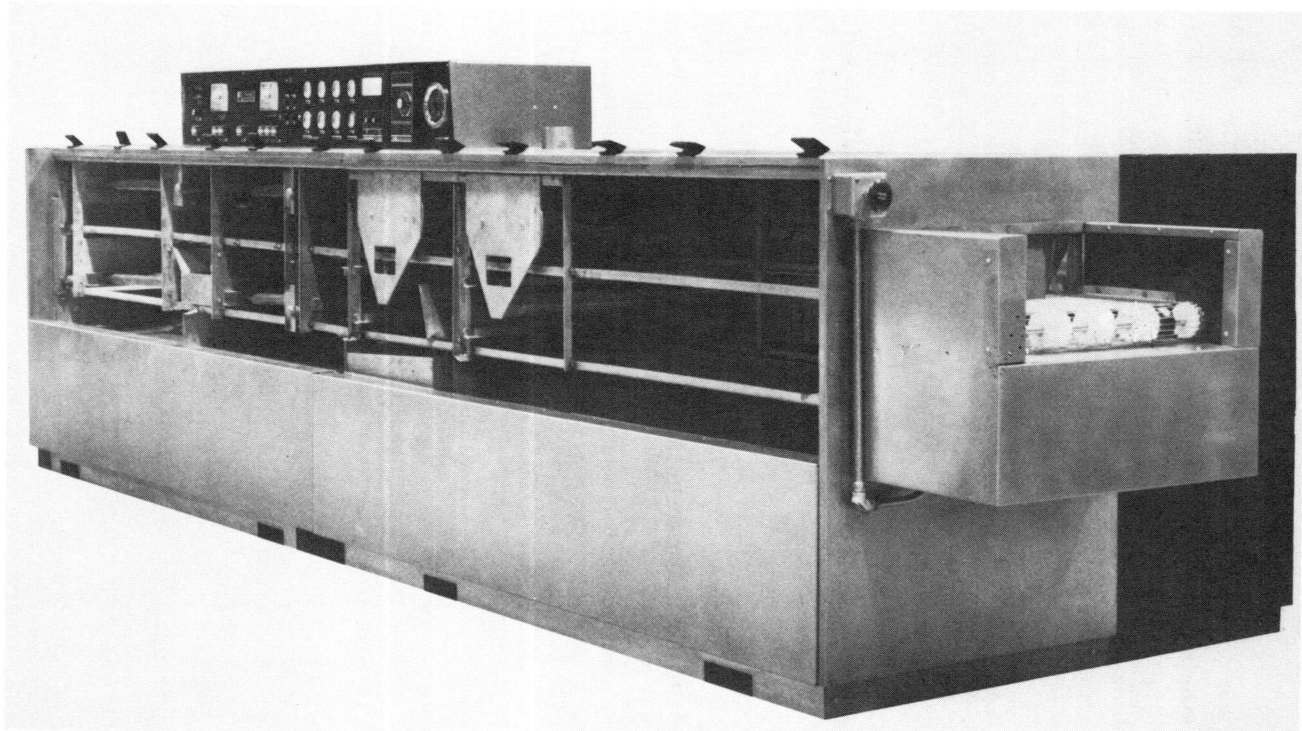

Figure 5.18 HK-85 showing open sections. (Courtesy of Unique Industries Inc.)

5. Recirculated rinse
6. Fresh rinse input
7. Air blowoff
8. Heated air blowoff
9. Air blowoff (option)
10. Heated air blowoff
11. Exit end

These are standard configurations for industrial in-line aqueous cleaners. Enhancements are usually offered as options but in reality as redundant equipment process sections.

Entrance The entrance section consists of an open conveyor section where the product is placed onto the belt. The entrance section should be equipped with an on/off switch with delay timer, timed to length and speed of the conveyor. This could be a light interrupt type, so that when boards pass through and interrupt the beam, the system is activated. This starts all pumps (sprays) and oven blower. It should also turn on fresh water input.

Another function allows the machine to idle when no product is being processed. When activated, it measures and counts the number of boards going through the system to reflect the maximum load of the system.

Prewash (nonrecirculating)

- Control of upper and lower nozzle pressures is manually set, and pressure gauges monitor manifold pressure.
- Both the control and monitoring functions should be automated. This can be accomplished by motorizing the gate valve on the incoming manifold line from each pump. Usually, the control is on the bottom manifold, allowing it to be shut off completely and sending maximum pressure to the top manifold.

The volume of water must be monitored only coming from each pump and going through each manifold. Correlation should exist between pressure and volume in each section due to potential clogging of the manifolds. When this occurs, the pressure increases and the volume decreases, necessitating shutdown of the system.

This type of section usually sends the effluent directly to the drain, or to a separate holding tank. If the effluent is sent to a holding tank:

- Monitor and semiautomatically control water levels in tank
- High-level discharge can be automatically controlled to drain or designed for continuous drainage.
- Note that this effluent may require adjustment, i.e., pH, BOD, COD, metal content, etc., prior to discharge to the municipal sewer or local holding tank. Customers should check with local requirements to certify compliance to local regulations.

Prewash (recirculated system)

- Upper and lower nozzle pressures are manually controlled.
  - High and low water level sensors are usually built into the system.
  - High-level controls shut off the water.
  - Low level controls the addition of water, to prevent motor burnout or, if heated, to prevent heaters from overheating.
- Both the control and monitoring functions should be automated.

Recirculated Wash

- Upper and lower manifold pressures are manually controlled.

- Wash temperature control (set point ±) (on/off) manually controlled.
- Water additions to measure the amount of water added to the tank are automated through the level sensors.
- Low-level (number 1) water sensor for heater activation.
- Low level (number 2) for pump activation.
- High level for water shutoff.
- Amount of detergent added is manually controlled.

Rinse to Drain

- Primary function is to rinse the residual contaminants from the wash tank that are still ont the surface of the board and entrapped beneath some of the components.
- The secondary function is to remove as much contamination as possible so as not to contaminate the recirculated rinse tank.
- This rinse is supplied by the recirculated rinse section and the pressure is generated by the recirculated rinse pump. In the majority of systems, it is a single-sided rinse manifold.
- Usually there is no pressure monitoring in this section.

Recirculated Rinse

- This is the third rinse section of the aqueous system. Spray manifolds are supplied water by the rinse pump, and the pressure is monitored by pressure gauges and controlled manually.
- In a detergent system, the intent of this section is to promote additional rinsing to remove the detergent contaminants from the surface of the boards and beneath the com-

ponents. The total amount of water sprayed is approximately equal to that sprayed in the recirculated wash tank.

- In a water system, the intent of this section is to enhance the cleaning action of the system. Typically, this rinse is cleaner than the preceding rinses and helps provide a cleaner product.
- Both the monitor and control funtions should be automated.

Freshwater rinse

- This is a makeup for the recirculated rinse tank.
- This is source water, regulated by existing local water pressure and controlled in volume by the size of the nozzles. The size of the orifice in the nozzles in conjunction with the water pressure determines the volume of water that is sprayed onto the product.
- Usually, there is no identification of pressure in this section; just a monitoring of gallons per minute incoming flow, pressure is regulated through nozzle configurations.

Some systems have extra nozzles in the freshwater rinse to augment the rinse and increase the water supply in the recirculated rinse tank. This is done to replenish the volume depleted by the outgoing drag-out rinse. It acts as a skimming function for the removal of foam in the detergent wash system.

Air Blowoff

- The theoretical concept is evolutionary: allow the water to evaporate off the surface of the module by moving hot air over its surface to accelerate the evaporation. Newer systems utilized increase the air pressure to mechanically remove the water.

Presently, all systems have self-generating high-pressure air blowoff units. Although they use various size blowers (e.g., 5, 7, 10 hp) and piping configurations, conceptually they are all similar.

Mechanically remove as much water as possible with the air knives. It has been found that this method of drying has the additional benefit of cleaning the boards more effectively. As the water is blown off the surface of the module, the contaminants suspended and dissolved in the water are removed as well, which results in a cleaner product.

Removing water by mechanical means is less energy-intensive than utilizing the evaporation method. The energy needed to evaporate water is 972 BUT/lb, and roughly calculated works out to be approximately 50 kw-hr. Therefore when designing dryer sections, higher values of kw-hr are utilized for safety purposes. However, to blow off the water the energy consumed by the motor and blower unit is substantially less, approximtely 4.0 kw-hr, and hence more efficient.

High-volume systems have dual air blowoff sections (see Fig. 5.19) to allow for higher throughput with faster conveyor speed. Also, usually incorporated in the blowoff sections are heated air generators, which are much lower in kw-hr than the standard evaporation-type dryers. The various types of heaters used for this purpose are:

- Quartz lamps
- Calrod heater coils
- Electric wire core heaters, etc.

In all cases, the heater function is to heat the air within the chambers and allow it to be pulled through the blower system, so that it can be used in the high-volume blow-off cycle. As the system heats up and the blowoff air is heated, drying efficiency is improved due to the capacity of warm air to hold more moisture.

Each oven section of any aqueous cleaner has switches for pump activation and temperature set/control units or equivalent for setting and maintaining oven temperatures. Manual adjust-

Figure 5.19 HK-85 dual air knife system. (Courtesy of Unique Industries Inc.)

ments are available for setting the air manifold height off the conveyor belt. These are usually difficult to adjust and are set "one time" for the product with the highest components.

The temperature in the oven is subject to many variations due to air flow controls within the section. Venting systems that are improperly adjusted will remove the heated air from the oven and reduce the system efficiency. Therefore, monitoring the entire chamber, including air flow and temperature of the outgoing gases in the vent system, is highly recommended.

Automatic controls and adjustment should be made available for air knife height above the conveyor belt. Fixturing and worm gear arrangements or pneumatic systems could be designed to accomplish this function. Air pressure should be monitored and controlled on the manifolds, either by adjusting the velocity through each air knife or having the ability to shut off certain air knives on demand.

The velocity through each knife can be adjusted by varying slot width in manifold pipes. The ability to shut off various air knives can be accomplished by either closing off the air slots or preventing total air from entering specific air pipe manifolds. See Fig. 5.20.

Effluent Control

Effluent discharge from aqueous cleaners is usually captured, monitored for pH, corrected, and discharged. The capturing is accomplished by tying the drain lines from prewashers, wash tanks, drag-out rinses, recirculated rinses, and any other optional rinses, to a supplementary tank attached to the system. Note: this tank must be sized to accommodate at minimum the volume of the wash tank. Also, it should be placed in a location where gravity can work with the system, in a beneficial manner to prevent overflow by creating an equilibrium condition between effluent tank and supply tank.

Continuous monitoring of pH is necessary in the effluent tank. As the acidity levels increase, a buffered solution must be added to neutralize the material or bring it to the requirements as defined by local community regulations.

5.5 SYSTEM AUTOMATION

Controls in effluent tank are as follows:

- Monitor and control high and low solution levels.
- Monitor and control pump on/off.
- Monitor and control auto drain system.
- Monitor and control pH adjustment.

Sections to control and monitor are as follows:

	Monitor	Control
1. *Entrance*		
On/off with delay	X	X
2. *Prewash*		
Manifold pressure		
Top	X	X
Bottom	X	X
Solution vol. (gpm)		X
3. *Prewash (Recir)*		
Manifold pressure		
Top	X	X
Bottom	X	X
Tank solution level		
Low	X	X
High	X	X
4. *Wash tank (detergent system)*		
Detergent add	X	X
Manifold pressure		
Top	X	X
Bottom	X	X
Tank solution level	X	X
Low	X	X
Intermediate	X	X
High	X	X
5. *Drag-out rinse*		
1st rinse		
Incomeing volume	X	X
Outgoing volume	X	X
6. *Recirculating rinse*		
Manifold pressure		
Top	X	X
Bottom	X	X
Tank solution level		
High	X	X
Low	X	X
Conductivity or pH pump volume		X
7. *Final rinse*		
Temperature		X
Incoming volume		X
Incoming pressure		X

	Monitor	Control
8. *Oven blowoff*		
Manifold pressure	X	X
Manifold height	X	X
Oven temperature	X	X
Manifold temperature	X	X
9. *Effluent tank*		
Incoming volume		X
Outgoing volume		X
Tank solution		
High	X	X
Low	X	X
pH	X	X

Notes:
1. Relationship must exist between volume and pressure because if pressure increases and volume remains constant, clogged nozzles are implied.
2. Both incoming and outgoing volumes must be equal; if not, this identifies an overflow condition.

5.5.1 Justification of Automation

Justifications of automated controls is difficult, especially since no one particular parameter can be economically reviewed to determine its financial value. To this end, when this type of equipment is implemented, as in prototype equipment or manufacturing equipment, the engineer should be made aware of the potential of itemizing all the process parameters. Once the itemization has occurred, then cost can be assigned and justification can be formulated.

For introducing the product into any manufacturing facility, certain assumptions have to be utilized to develop "gut" feelings as to what this system is worth. The worth of the system must be determined by the need and ability to quantify the process used by the buyers. Following are some of the elements that must be addressed in justifying an automated system:

Figure 5.20 Hollis air knife. (Courtesy of Hollis Automation Inc.)

- ○ Ease of conducting experiment
- ○ Ability to control all elements and varying predetermined elements
- ○ Ease of collecting data
- ○ Savings of approximately 50% of the time needed to collect data
- ○ Operation
    - ○ Repeatability of process
    - ○ Optimization of process
        - ○ Improves product quality
        - ○ Improves process quality
        - ○ Reduction in defects minimum of 25%
        - ○ Improves process troubleshooting by minimum of 50%
        - ○ Improves throughput by minimum of 15%
- ○ Automatic controls
    - ○ More efficient use of equipment
        - ○ Actual knowledge of on/off cycle
        - ○ Better predictability of downtime and maintenance (or preventive maintenance)
- ○ Consistency or repeability of process
        - ○ Reduces problems by 50%
        - ○ Inconsistent process affects the following:
            - ○ Conveyor speed
            - ○ Wash temperature
            - ○ Oven temperature

- Air flow
- Pressure of nozzles
- Venting
- Wash detergent concentration
- Conductivity of incoming water

- Energy efficiency
  - Active system only when needed
  - Saves water
  - Can idle the heaters, saving kw-hours.

5.5.2 Justifying a Cleaning System

Once the decision has been made to clean and the solvent has been chosen, then deciding which equipment to get, batch or in-line, must be based on sound financial cost and engineering evaluations. Capacity of equipment, cost of solvent usage, and disposal of spent solvent are some of the elements used in the justification of any system.

Justification Consideration

The following is an example of a procedural justification which includes not only equipment considerations but considers environmental impact as well.

Products known

1. Size of product ___ X ___.
2. Volume of product ___.
3. Projected volume ___.

Facilities

1. Power available ___ Voltages ___ Amps

2. Plant compressed air ___ psi
3. Water capacity ___ gpm
4. Water temperature ___ °F (°C)
5. Drains ___ pipe diameter
6. Source of disposal

   Sanitary sewer ___

   Septic system ___

   Local water access ___
7. Certified disposal company ___
8. State disposal requirements ___
9. Federal disposal requirements ___

5.5.3 Equipment

To define which size equipment or system to purchase, proceed to answer the following statements:

Solvent Batch Degreaser

1. Sized for product (opening area = square area)
2. Size of cooling coils
3. Number of rinse dips
4. Depth of boiling sump
5. Availability of spray wand
6. Heaters sized for solvent
7. Freeboard area
8. Distillation rate
9. High-temperature control

10. Automatic shutdown
11. Still (if available)
12. Materials of construction (stainless steel)
13. Dryer setup for water removal from solvent
14. Automatic crane to convey the product through the system
15. Venting requirements (CFM)
16. Amount of drag-out evaporation

Solvent In-Line System

It is necessary to understand the impact of the following nine elements when deciding on which in-line system to purchase.

1. Heating capacity
2. Solvent capacity
3. Distillation rate
4. Spray nozzles:

   Quantity above and below belt

   Height off the belt

   Spray pressure

   Spray volume
5. Conveyor belt width
6. Angle of conveyor going into and out of chamber
7. Solvent usage and volume evaporation capacity
8. Cooling capacity of chiller system
9. Venting system

Aqueous Batch System

Review and answer the following statements to determine which system to purchase:

1. Number of cycles needed, as defined per experimentation.
2. Type of water necessary:

   Softened

   Deionized

   Tap
3. Dryer cycle and number of drying cycles
4. Dryer efficiency
5. Rinse cycles
6. Capability of detergent add
7. Volume of water in each cycle
8. Availability of local drain
9. No local drain; ability to pump to remote location
10. If detergent is used, biodegradability (BOD) requirements, as outlined per local requirements
11. Prewash capability if using water-soluble flux

Aqueous In-Line System

In determining the size of system to purchase, a thorough knowledge of the process is mandatory. Review each element and develop your understanding:

1. Prewash necessity
   - Fresh water to drain
   - Recirculated rinse supply to drain

- Power recirculated
- Pressure of nozzles
- Volume of water

2. Wash
   - Length of section
   - Pressure regulators on upper and lower manifolds
   - Pressure of sprays
   - Coverage on belt (spray coverage)
   - Overall height available within system
   - Ability to maintain water height
   - Heating capability (with temperature controls)
   - Short heat-up time
3. Rinse
   - Sufficient spacing between wash and rinse tank, to prevent cross-contamination
   - Supplied by fresh incoming water
   - Supplied by recirculated rinse tank
   - Drain capability
   - Known volume of water
   - Known pressure
4. Recirculated rinse
   - Quantity of recirculated rinses required
   - Length of spray sections
   - Makeup capabilities if supply is used for initial rinse
   - Pressure gauges for upper and lower manifolds

5. Final rinse (dual function)
    - Makeup for recirculated rinse
    - Final freshwater rinse
    - Amount of spray
    - Length of spray
    - Coverage

Between all wet sections, diffusion barriers must be installed to prevent cross-contamination of the tanks. These barriers could be rubberized curtains or air knife curtains, where any cross-spray would get knocked down into its respective tank.

6. Oven blowoff section
    - Self-contained air blowoff
    - Quantity of dryer sections
    - Height of air knife above the belt
    - Quantity of knives above and below belt
    - Horsepower of blower units
    - Length of dryer sections
    - Width of orifice of air knives
    - Design to get equal pressure across the entire belt
    - Section on bottom for catching all blown off water
    - Drain for blown off water
    - Curtains within the section to prevent escapting pressure into other chambers
7. Heated blowoff section
    - Hot-air generator coils, I/R, etc.
    - Blowoff section, with air intake potentially capturing hot-air system

8. Overall system design

   - Easy access to all sections
   - Maintenance easily accomplished
   - Pump and motors accessible
   - Sensors easily reached and cleaned
   - Belt easily changed with more than two break points, recommended three
   - Belt materials durable
     - Stainless Steel

Note: Consider the slipperiness of belt if it is angled; boards will slide one upon the other which would be detrimental to the components.

9. Operational

   - Sensors at entrance, activating pumps and sprays for given time period
   - Idle temperatures for all wash heaters

5.6 TESTING FOR CLEANLINESS

There are many ways and various types of equipment to check for cleanliness (see Figs. 5.21 and 5.22). Mil-P-28809 is the specification that brought everything to the forefront. Prior to that time, everyone had an idea but no one had a number to quantify the results. Now that the results can be quantified (see Fig. 5.23), limits have been placed on acceptable and rejectable levels. The levels have never proven electrical failures; they have only identified poor adhesion of conformal coatings to the laminate materials.

The tests are also designed to measure the residual rosin materials left on the board after the cleaning process. The solution in the solvent extract testers are a mixture of deionized water and isopropyl alcohol to a mixture of 75/25%, respectively.

Figure 5.21 Kenco Omega Meter Model 600. (Courtesy of Kenco Industries Inc.)

Figure 5.22 Alpha ionograph. (Courtesy of Alpha Metals Inc.)

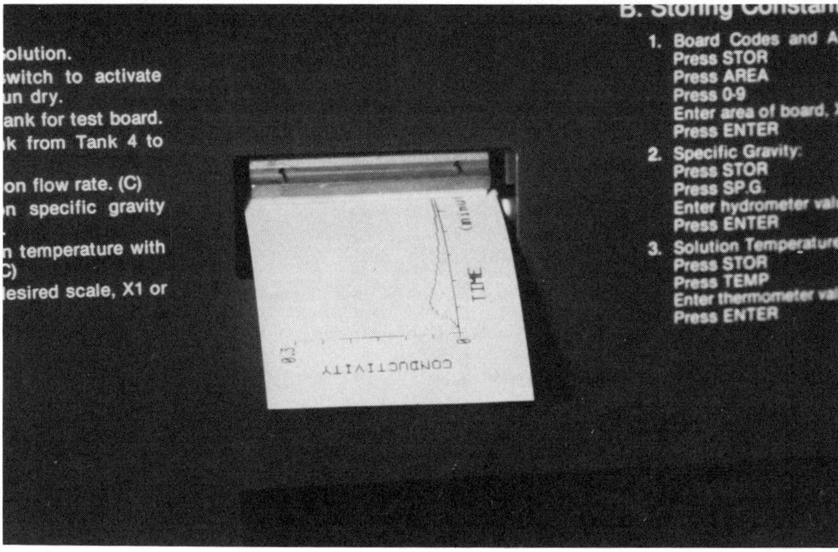

Figure 5.23 Alpha Ionograph chart recording. (Courtesy of Alpha Metals Inc.)

Reemphasizing, the solution was designed to dissolve the rosin and measure the residual ionics.

5.7 SAFETY REQUIREMENTS

The use of these materials must also be monitored very closely due to Environmental Protection Agency (EPA) and Occupational Safety and Health Association (OSHA) requirements. The threshold limit value (TLV), or the exposure level that humans can be exposed to safely for a given period of time, is the gist of how these materials will be handled in a manufacturing cleaning operation.

Proper venting requirements are guided by the ability to remove the fumes from the areas where the degreasing operation is occurring. A note of caution: fumes from solvent are typically heavier than air and venting installations must be installed on the equipment, drawing the material or fumes horizontally away from the degreaser.

5.8 GENERAL DEGREASER EQUIPMENT DESIGN GUIDELINES

To prevent fumes escaping from any type of degreaser, cooling coils are mounted around the periphery of the inner walls of the degreaser. These tubes are cooled with either a self-contained coolant cooled through a chiller or tap water with a temperature low enough to have the hot vapors condensed on its coils. Some systems have a double set of cooling coils to enhance the condensing effects of the boiling solvent. The various sections of the vapor degreaser and their functions are discussed below.

5.8.1 Sizing of Mechanical System

Size of opening is dependent on the size of boards that will be processed. A board of 10 in. X 12 in. would require a minimum tank width of 10 in. plus the width of the basket that will be used to suspend the boards within the system. The depth of the tank is determined by whether or not the boards will be submerged during the cleaning cycle. If the answer is to submerge, then the sump tank must be deep enough to accommodate the board.

Size of heaters is determined by the type of solvent being used. Caution must be exercized to certify the correct heaters. The watt density capacity of the heaters must be known and accurate for each solvent type. Solvent types should not be changed without a thorough understanding of the process ramifications.

If the heaters are too hot, the solvent will break down and deteriorate. When the solvent breaks down, the possibility of acid generation is high, and this could eventually corrode the degreaser, rendering it useless.

Some degreasers don't use heaters but utilize a heat exchanger mechanism from the cooling process of the chiller coils. This is a worthwhile concept if the proper solvent is chosen.

Cooling coils must be made of stainless steel and have enough surface area to cool the solvent vapors. The purpose of the cooling coils is to prevent the vapors from escaping the tank of the degreaser. The coils accomplish this by being cool enough to condense the vapors. However, the coils must have enough capacity to always maintain a stable vapor height, and not overpower the vapors to lower them to a point beyond which they may not be high enough to perform their cleaning function. Different thermostatic controls are usually incorporated into degreaser designs to accomplish these functions.

High-temperature vapor sensors allow for design to increase the volume of coolant through the coils, while the low-temperature sensors allow for increased capacity of heat being applied to boiling sump.

5.8.2 Recovery System

Another function of vapor degreasers is the ability of the system to generate fresh solvents. All the vapors condensed off the cooling coils are captured and returned to a clean holding tank, which could be internal or external to the degreaser. This is important if a spray wand is utilized to mechanically spray the product within the tanks. If the system cannot generate enough to maintain the need of the spray wand, then it becomes a useless option.

Freeboard, the distance from the top cooling coil to the entrance surface of the degreaser, is also a critical dimension. It is recommended that this distance or dimension be 75% of the longest opening dimension of the degreaser. The purpose of the large freeboard is to prevent the fumes from escaping during the operation of the system. For example, for a degreaser with a 1 ft × 2 ft opening, the freeboard should be 1.75 ft, or 75% of the 2-ft dimension, from the top cooling coil.

Materials for degreaser construction should be stainless steel, including the cooling coils and welds. This will prevent the decompositional products of the solvents from attacking the degreaser.

5.9 SOLVENT EFFECTIVITY LEVELS

The following subsections are used to determine the effectivity of solvents used for cleaning the residual fluxes off the printed wiring boards.

5.9.1 Solvent Extract Test

The boards are placed in a given quantity of alcohol and deionized water, which has been brought to a given level of cleanliness, as defined by a resistivity measurement called megohms. As the contamination is dissolved from the board, the resistivity of the solution changes, defining the amount of contamination removed from the product. This change is analogous to an amount of sodium chloride per given area of the product tested. This is compared to the requirements of MIL-P-28809.

5.9.2 Acidity Test

This test is to check for acidity levels of the solvents. When a given amount of water is absorbed through either the condensation from the cooling coils or high humidity levels in the area, a reaction occurs with the chlorides in the solvents, creating an acid. Depending on how much moisture is absorbed and the acid acceptors left in the solvent, the acidity levels will vary to such an extent that defining the proper level is impossible.

5.9.3 pH Acid Acceptance Test

Recent work has shown that pH measurements are unreliable and acid acceptance (as wt% of NaOH) is more reliable and repeatable. This work has been correlated to the amount of corrosion that occurs on the tin-lead plating after being subjected to various levels of acid in the solvent. "Acid acceptance testing measures the amount of free (strong) acid that a solvent can absorb with no damage to either the solvent or the printed circuit boards" [1].

In conclusion, there are a variety of methods for checking solvent efficiency as previously mentioned. However, each method must be individually analyzed to determine the most efficient process for your product. Work performed by the author has been with solvent extract test and acidity measurements. The acid measurement was never found to be consistent, as related to the vendor's specification; therefore, with the new surface mount components that are now being used, cleanliness becomes critically important. Hence, the author recommends the acid acceptance testing procedures described above to properly determine the effectivity of any and all cleaning solvents.

Once identified as bad material, the substance should be neutralized and reclaimed by the vendor. Proper disposal of any and all solvents must be carefully planned and executed. Disposal companies must be licensed by the state or federal government. The Resource Conservation and Recovery Act (RECRA) [2] requires all material to be documented and controlled from fabrication to disposal. Users of any hazardous materials should realize that good work practices protect the environment and make their company a better neighbor.

BIBLIOGRAPHY

1. Jeanne Giroux, Steven Nowak, *The Unimportance of pH*, Circuits Manufacturing, September 1982, pp. 36-38.

2. Resource Conservation and Recovery Act, Federal Register, November 17, 1981.

ASSOCIATED READINGS

1. Carl J. Tautscher, *Terminology Used in Conjunction with Electrical Insulation Resistance Testing of Printed Wiring*, IPC Proceedings, Fall 1982, pp 1-9.

2. Dr. Jack Brous, Alvin F. Schneider, *Cleaning Surface Mounted Assemblies with Azeotropic Solvent Mixtures*, Electri-Onics, April 1984, pp 54-55.

3. Jeanne Giroux, Steve Nowak, *The Unimportance of pH*, Circuits Manufacturing, September 1982, pp 36-38.

4. Alvin Kaufman, *Evaluating Cleaning Efficiency with Radioactive Isotopes*, Circuits Manufacturing, May 1976, pp 34-37.

5. William A. Hume, *Proper Cleaning of Electronic Assemblies*, Assembly Engineering, March 1968, pp 37-41.

6. W. B. Wargotz, *Cleanliness and Reliability*, Circuits Manufacturing, February 1978, pp 40-49.

7. *Silver Nitrate Test–A Simple Cleanliness Test*, Alpha Metals Inc., Bulletin TR-2000.

8. *How Clean Is Clean?* Quality Management and Engineering, May 1974, pp 42-45.

9. Howard H. Manko, *How Clean Is "Clean" in Electronics?* Quality Assurance, June 1966.

10. *Bubble Cleaning System Removes Contaminants from Core Memory Assemblies*, Assembly Brief, Assembly Engineering, October 1978, p. 28.

11. Harry C. Glover, *Are Emulsified Solvents Safer Cleaners?* Production Engineering, July 1978, pp 41-48.

12. J. Brous, *Evaluation of Post Solder Flux Removal*, Welding Research Supplement, December 1975.

13. Donald A. Elliott, *Modern Pollution-Free Cleaning Methods for Printed Wiring Assemblies*, publication data and place unknown.

14. H. H. Manko, *Beware of Electronic Dirt*, Electronic Design, November 1969.

15. *Solvent Compatibility and Electronic Cleaning*, Electronic Packaging and Production, April 1971.

16. Allen Elzback, *Cleaning Equipment Survey*, Electronic Packaging and Production, March 1974.

17. William G. Kenyon, *Improved RA Flux Cleaning with New Fluorocarbon/Alcohol Combinations*, Nepcon 77 Proceedings.

18. Leo Lambert, *Aqueous Cleaning and Why at Digital Equipment Corp*, Nepcon West Proceedings, 1978.

19. Department of the Navy, *Materials Research Report No. 13-77*, Naval Avionics Center, November 1977.

20. Donna Sager and Kathryn Johnson, *A Study of Solvent and Aqueous Cleaning of Fluxes*, NWC-TP-6427, Naval Weapons Center, February 1983.

6

Manual Soldering

6.0 MANUAL SOLDERING

Manual soldering is the oldest method of soldering. The process consists of bringing solder to the parts being joined and applying heat through an object called a solder iron. History reveals that the solder irons were once heated by fire and were brought to the work in a red hot condition. See Fig. 6.1 for different styles of irons.

Manual soldering is a very complex and misunderstood operation. Soldering iron tip temperatures reach 800°F and when misuse occurs or when an operator improperly uses the solder iron, the reliability of the solder joint and the product becomes suspect and the quality becomes inconsistent. This inconsistency from joint to joint creates potentially defective product.

Through the years improvements have augmented the efficiency of the solder iron, all of which have been driven by technological advances of the products to be soldered. In the 1950s and 1960s the solder iron improved from the standard two-wire cord to the three-wire cord, which included a ground termination wire. Improvements were also made in the cord so that it would not melt if it came into contact with the iron itself as shown in Fig. 6.2. During the 1970s and 1980s major improvement were again made in the temperature controls to improve the consistency of the tip temperature. Enhancements were again made for the prevention of electrostatic discharge and voltage spikes which occurred when the tip was placed on the work. Tip developments were seen during this period. Copper tips needing to be filled and shaped were replaced with new iron-plated and tinned-iron tips. The new tips come in all configurations of sizes and shank sizes as shown in Figure 6.3, thereby allowing the iron to be properly chosen for the soldering applications required.

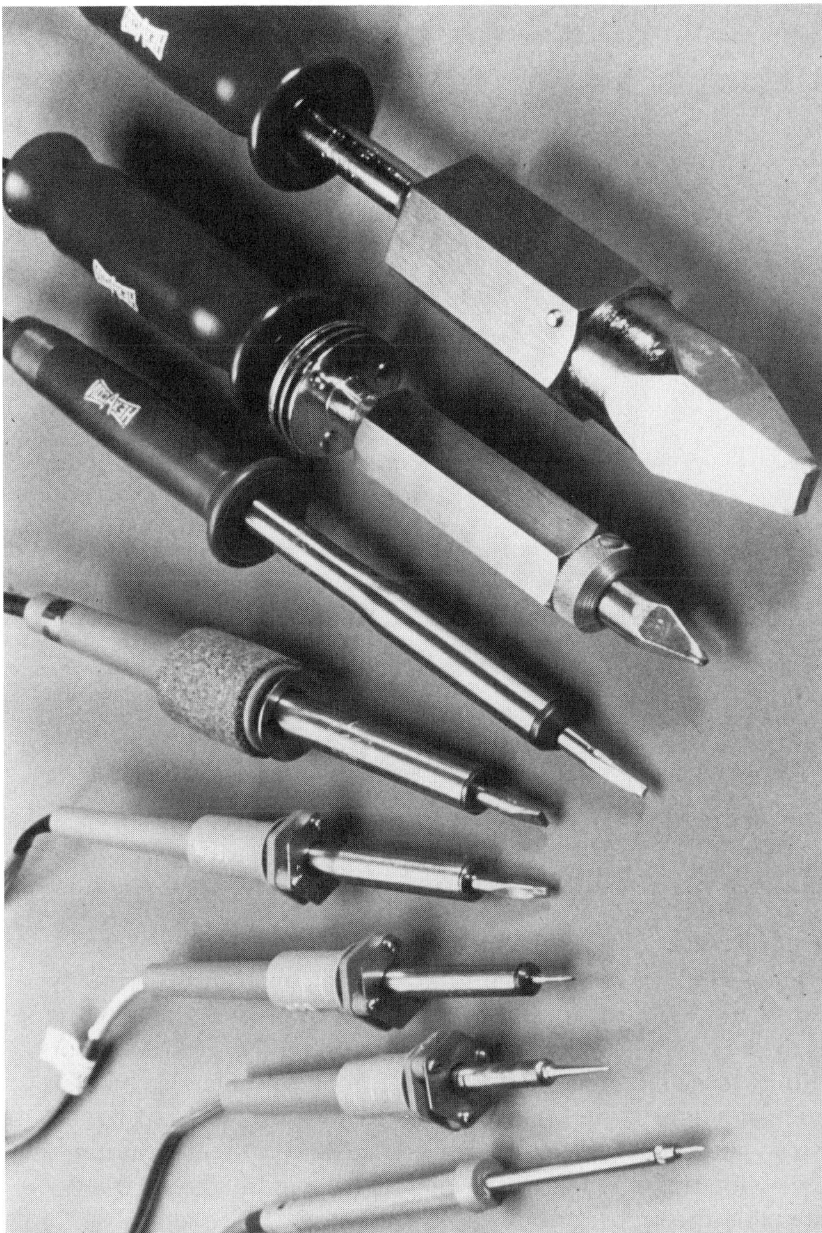

Figure 6.1 Various size soldering irons. (Courtesy of Hexacon Electric Company, Roselle Park, NJ.)

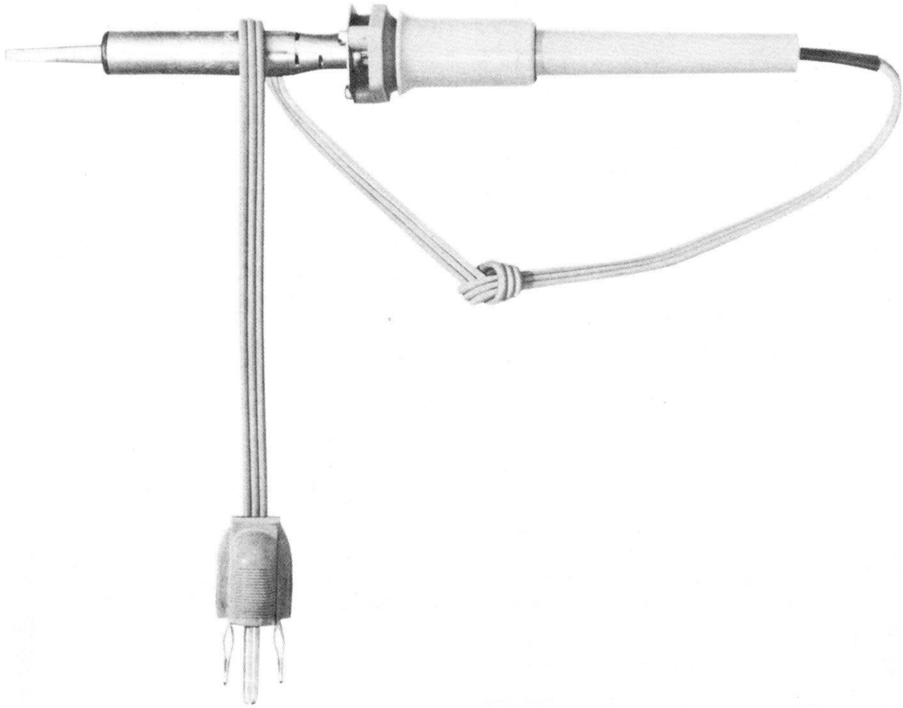

Figure 6.2 High-temp. Super-flex cord. (Courtesy of Hexacon Electric Company, Roselle Park, NJ.)

6.1 HOW TO CHOOSE A SOLDER IRON

The wrong choice of a solder iron will result in damage to the printed wiring board and components, or if the iron is thermally undersized it will not supply enough heat to properly reflow the solder joint. In choosing a solder iron, review the size of the solder joint to be created and perform temperature profiles of the particular solder joints. This will identify whether or not the solder iron and tip are of the proper voltage and tip temperature. It will also identify the recovery time of the soldering iron, which will enable the recovery rate to be determined. The analysis of

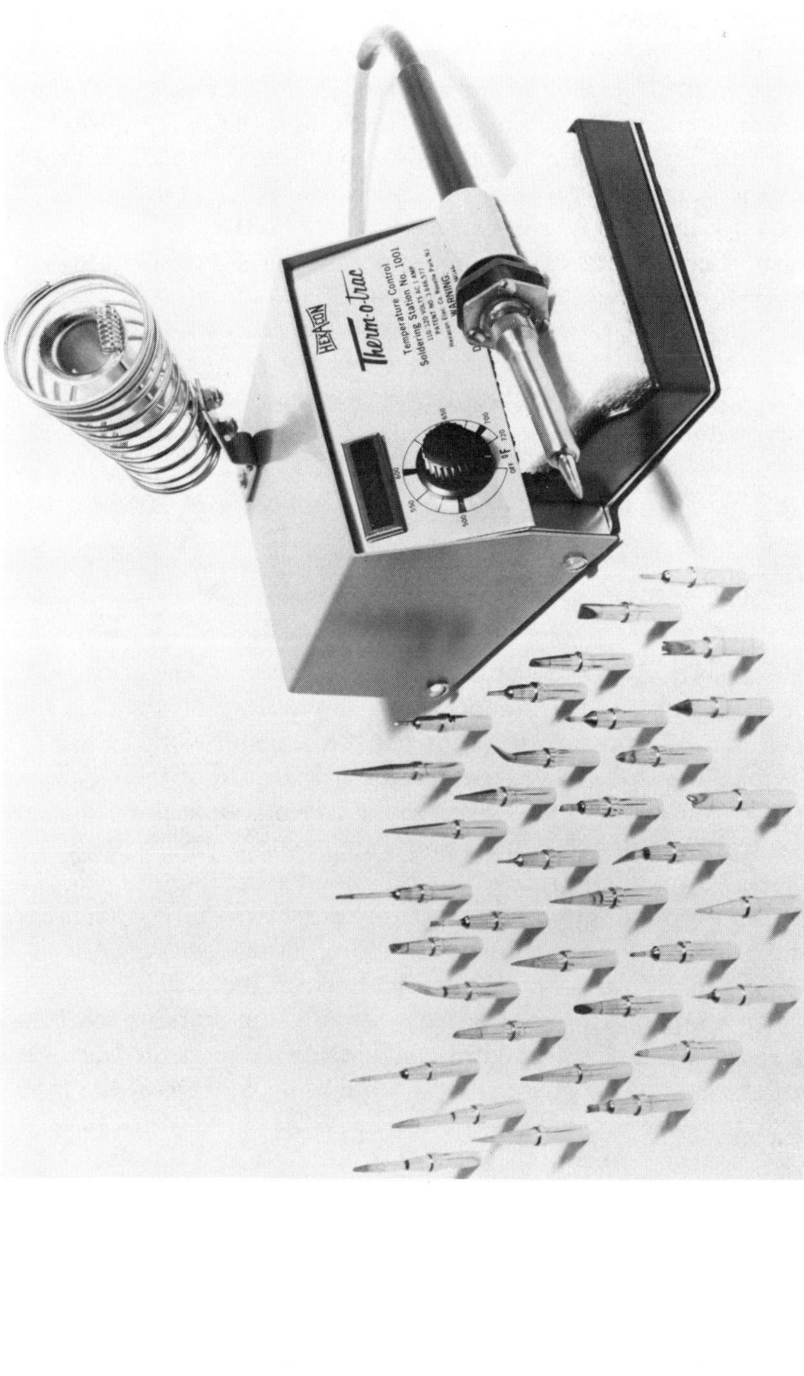

Figure 6.3 Variety of Soldering tips. (Courtesy of Hexacon Electric Company, Roselle Park, NJ.)

this information will determine which soldering iron is best suited to perform the solder joint.

Different size solder joints will require specific qualification of different solder irons. This qualification can be accomplished by using a temperature graphic analyzer (Fig. 6.4), which characterizes the temperature variation experienced when creating the individual solder joints as explained in Sec. 6.1.1.

Another phase in choosing a soldering iron is consideration of the operators who will be using the tool. The iron must be comfortable to use and be properly balanced for ease of use over the period of the day. It should not have any hot spots to burn the operators, and it must have a solder iron holder which will hold the tool without falling out. See Fig. 6.5 for solder iron with holder.

6.1.1 Calibration of Solder Irons

Special analytical tools must be used to calibrate solder irons. The major element is the thermal profile of the iron as it is creating the joint (Fig. 6.6). This profile identifies the idle and working temperature of the iron. If the temperature difference of the two is large, the iron will have to work too hard to maintain the proper temperature and as subsequent joints are created, the iron will be unable to thermally recover thereby increasing the time to create the joint and potentially creating defective joints or damaging the printed wiring assembly. The smallest temperature difference is the optimum for the best soldering results.

The temperature variation between idle and working temperature zones can be controlled by either the tip size or the wattage size of the iron. This attests that although the wattage of the iron is high enough to create the required heat, the tip may not have enough mass size to contain the caloric content of the wattage supplied by the iron. Undersized tips cool off quicker, and therefore more time is needed between successive solder joints to regain the tip temperature required for another solder joint.

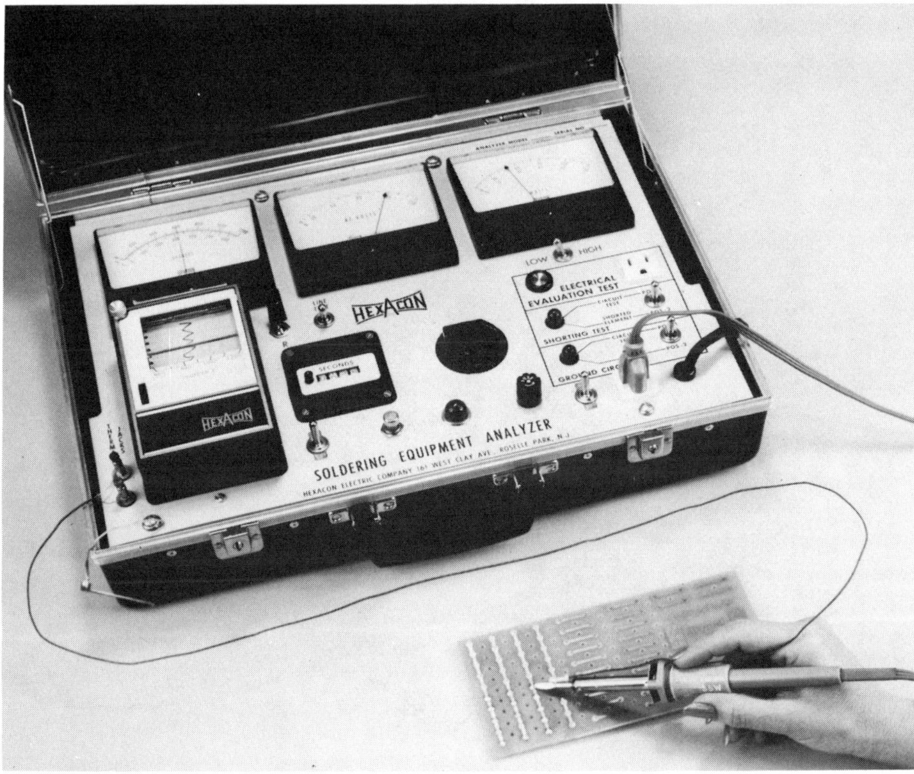

Figure 6.4 Soldering equipment analyzer. (Courtesy of Hexacon Electric Company, Roselle Park, NJ.)

6.2 MANUAL SOLDERING TRAINING PROCESS

Should the length of training programs be dependent on skills needed to perform the job without destroying the product? Emphatically, yes! Plus more. Train the operators to understand the quality requirements and have the skills to make the product consistent with those requirements, as shown in Fig. 6.7. There are many professional organizations who supply these types of programs and adjust them for particular application or product categories. Seek these people; they can help. Figure 6.8 identifies the training triangle necessary to reach all levels of

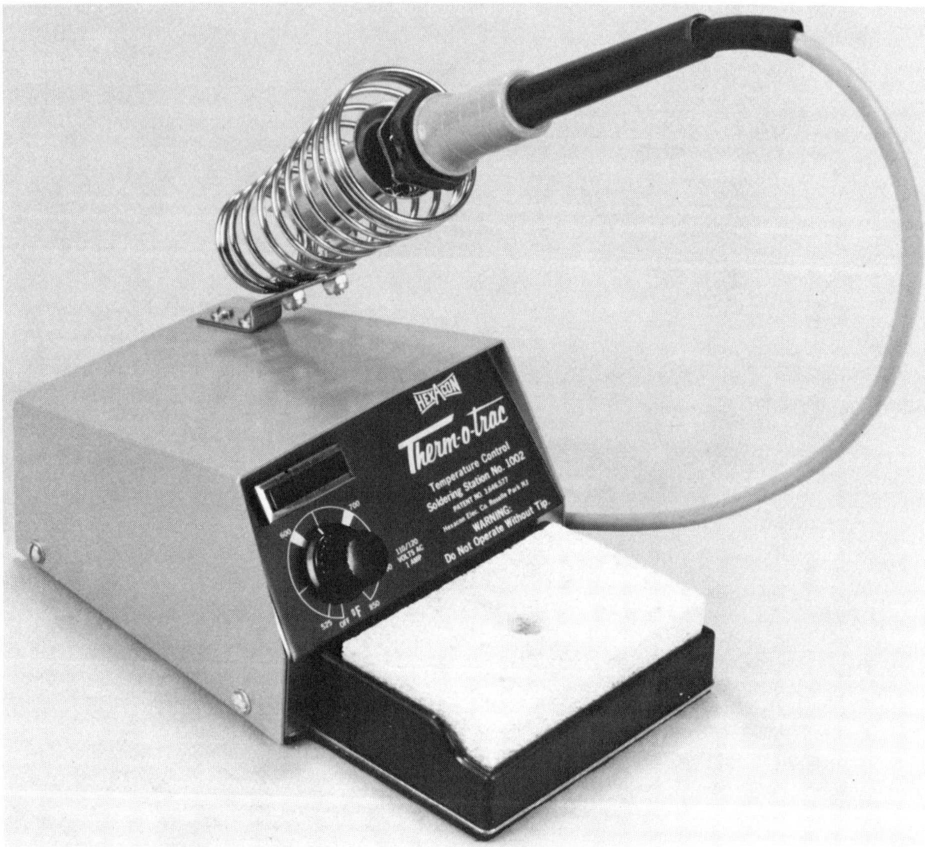

Figure 6.5 Therm-o-Trac in holder. (Courtesy of Hexacon Electric Company, Roselle Park, NJ.)

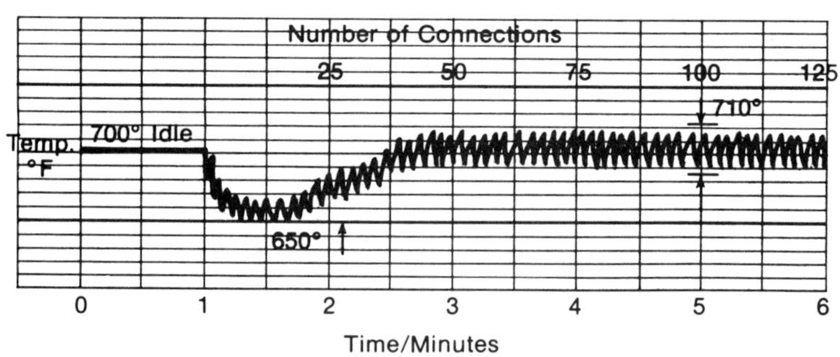

Figure 6.6 Graph of thermal profile. (Courtesy of Hexacon Electric Company, Roselle Park, NJ.)

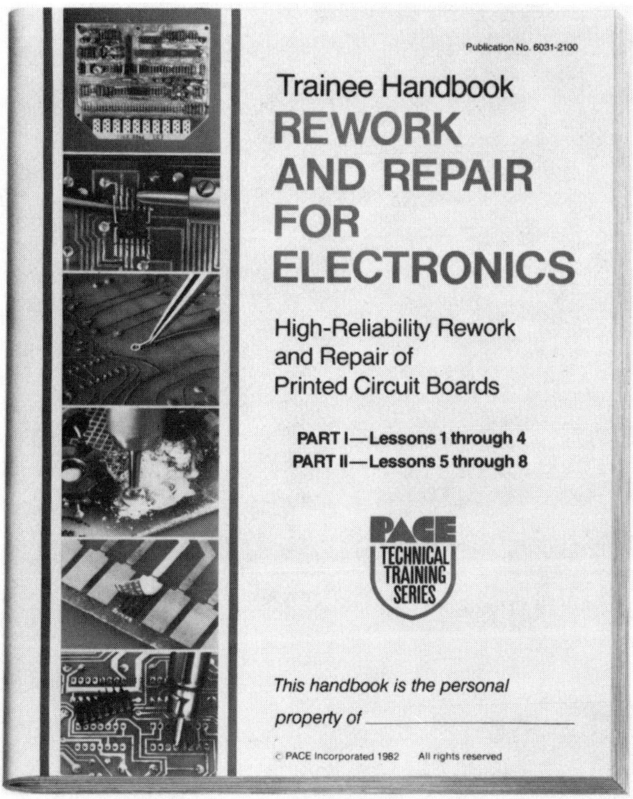

Figure 6.7 Trainee handbook, rework and repair for electronic. (Courtesy of Pace Incorporated, Laurel, MD.)

management for successful training program implementation. The process of manual soldering must include proper training of the operators. All training programs will sustain themselves financially if administered properly and continuously maintained. As once stated, "We cannot build tomorrow's product with yesterday's skills," which is precisely what is occurring when operators are not given the knowledge of methods and procedures. They will not develop the skills required to correctly perform the job.

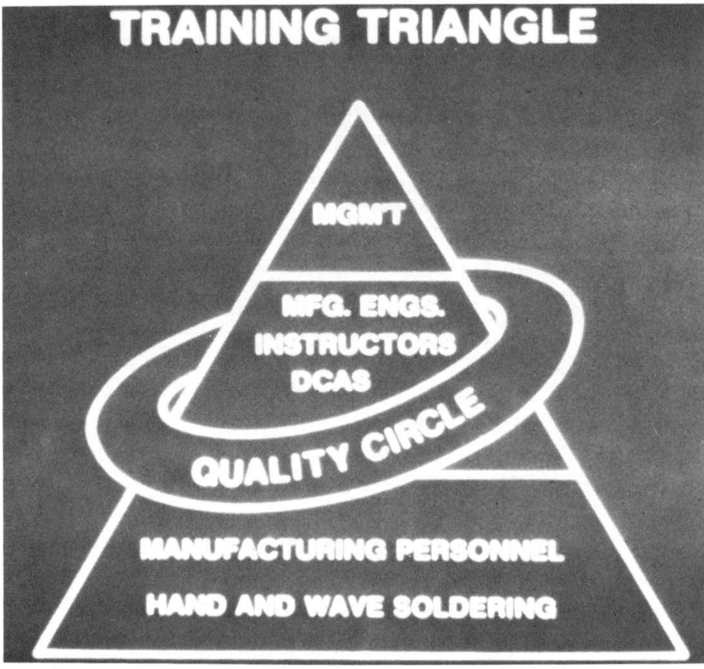

Figure 6.8 Training triangle. (Courtesy of OMNI Training Corp., Cucamonga, CA.)

Figure 6.9 Instructors guide. (Courtesy of Pace Incorporated, Laurel, MD.)

Training programs must contain both theory (Fig. 6.9) and actual work practices to achieve the skills required.

Special programs for operator training and certification must be seriously considered for all functional soldering operations. Business economics must dictate that no one without proper training and qualifications be allowed to perform critical soldering work. To allow this implies improper planning of producibility and manufacturability objectives. The golden rule should be "Don't allow ignorance to ruin your product" or "Ignorance of process results in scrap of product."

6.2.1 Training Program

Start with the basics:

- Describe the use of the solder irons, care, and handling.
- Describe product cost and importance of quality products:
  - Customer satisfaction
  - Operator pride
- Perform quality soldering functions on simulated products.
- Review quality and difficulties of achieving such.
- Review all repair tools the operators will encounter in the manufacturing process (Figs. 6.10-6.12).
- Have the operators practice with all the equipment on all the product types.
- Familiarize the operators with all pertinent documentation (Fig. 6.13).
- Identify steps to be taken when uncertain about procedures, equipment, or product. (Note: This is critical. Whenever a doubt exists, the repair must not be made.)
- Follow up on each individual once they are in manufacturing (Fig. 6.14).

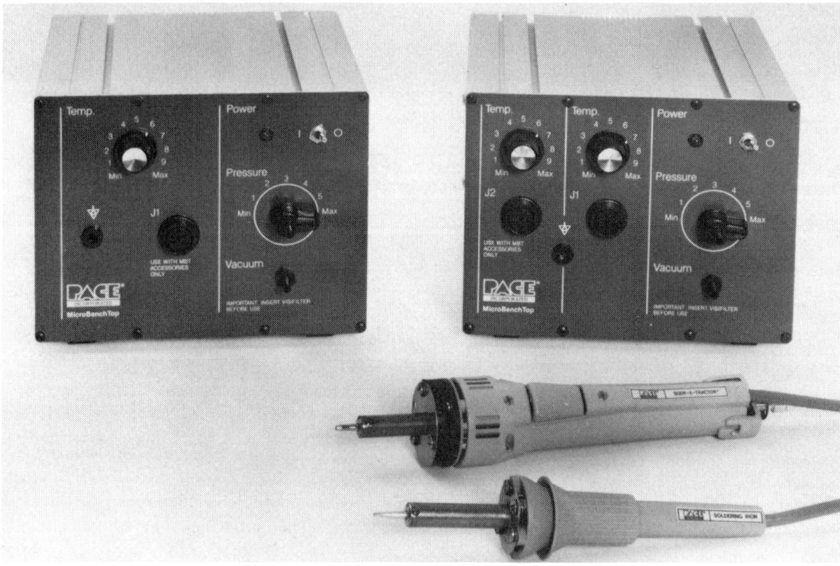

Figure 6.10 Pace MBT 100 and 200. (Courtesy of Pace Incorporated, Laurel, MD.)

Figure 6.11 Pace PRC 151. (Courtesy of Pace Incorporated, Laurel, MD.)

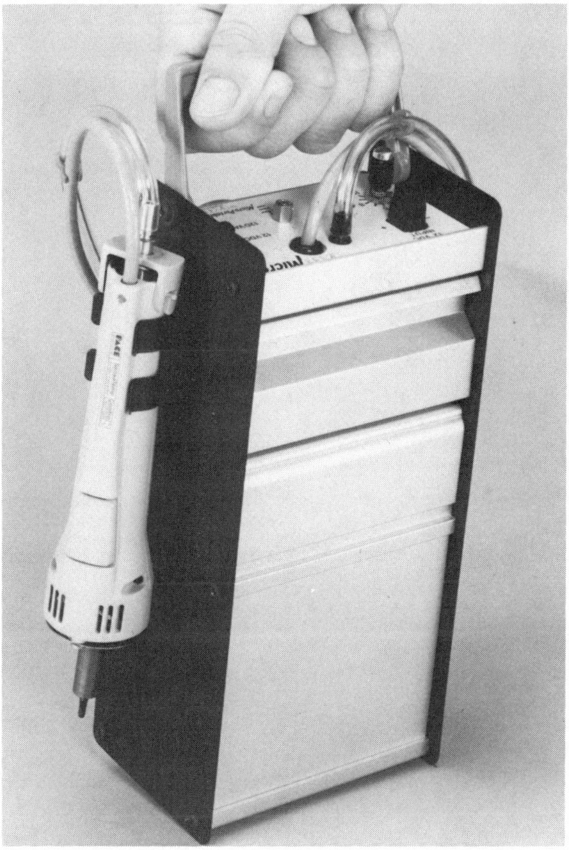

Figure 6.12 Pace MP-1 microportable. (Courtesy of Pace Incorporated, Laurel, MD.)

- Keep the training program current with equipment and process.
- Perform periodic retraining cycles on all operators (Fig. 6.15).
- Don't allow on-the-job training.

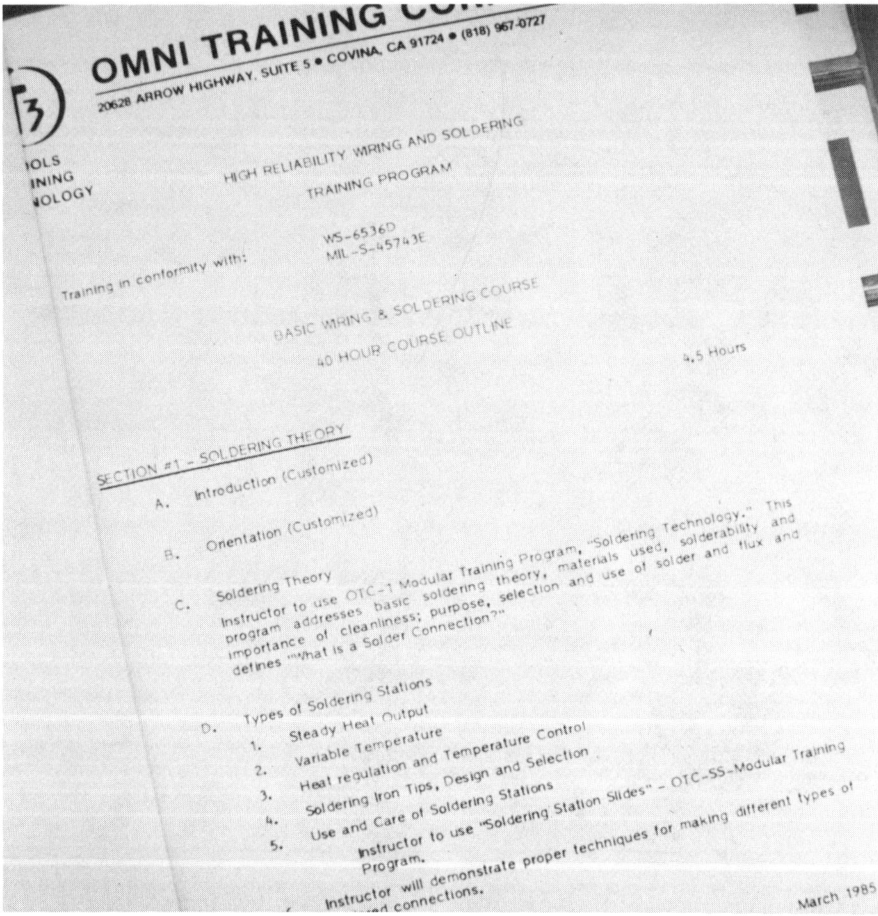

OMNI TRAINING CO...
20628 ARROW HIGHWAY, SUITE 5 • COVINA, CA 91724 • (818) 967-0727

HIGH RELIABILITY WIRING AND SOLDERING
TRAINING PROGRAM

Training in conformity with: WS-6536D
MIL-S-45743E

BASIC WIRING & SOLDERING COURSE
40 HOUR COURSE OUTLINE

SECTION #1 - SOLDERING THEORY 4.5 Hours

A. Introduction (Customized)

B. Orientation (Customized)

C. Soldering Theory
Instructor to use OTC-1 Modular Training Program, "Soldering Technology." This program addresses basic soldering theory, materials used, solderability and importance of cleanliness; purpose, selection and use of solder and flux and defines "What is a Solder Connection?"

D. Types of Soldering Stations
1. Steady Heat Output
2. Variable Temperature
3. Heat regulation and Temperature Control
4. Soldering Iron Tips, Design and Selection
5. Use and Care of Soldering Stations
Instructor to use "Soldering Station Slides" - OTC-SS-Modular Training Program.
Instructor will demonstrate proper techniques for making different types of ... connections.

March 1985

Figure 6.13 Book on training. (Courtesy of OMNI Training Corp., Cucamonga, CA.)

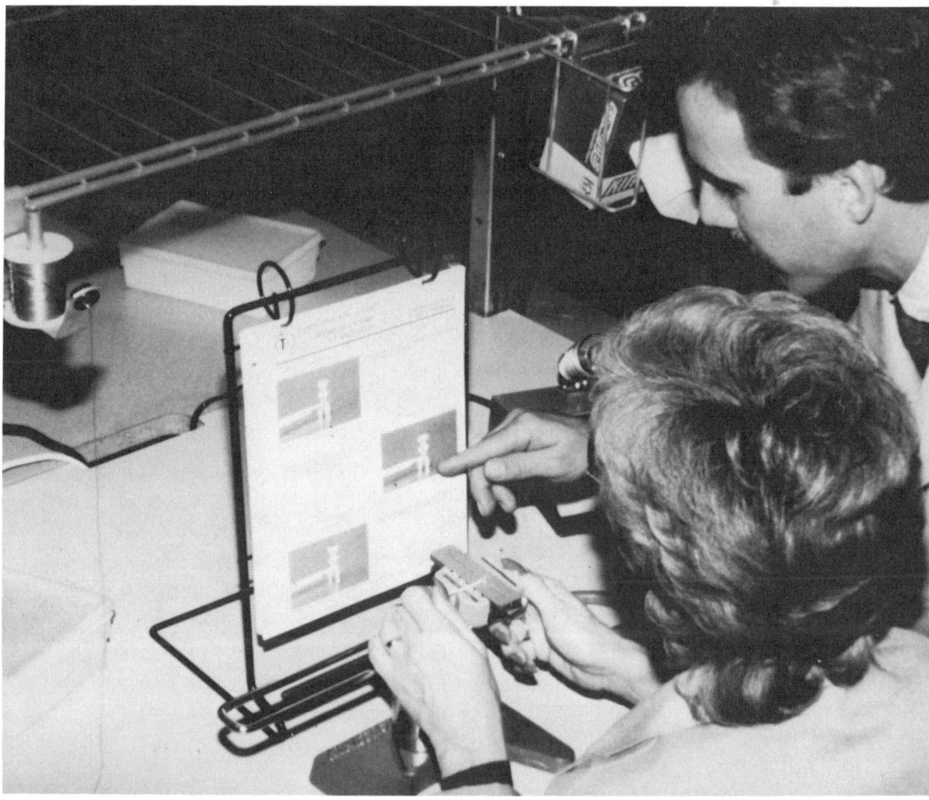

Figure 6.14 Operator and trainer. (Courtesy of OMNI Training Corp., Cucamonga, CA.)

6.2.2 Who Should be Trained?

Although many hours can be spent training the operators, don't forget the need to train the supervisors, engineers and quality personnel. Production supervisors are the support mechanism which must be used to certify the training is effective. They must also be knowledgable of what is being taught, to prevent ambiguity regarding interpretation of the standards. There is nothing more frustrating and costly as having a supervisor who does not agree with the training program, in both content and results, and

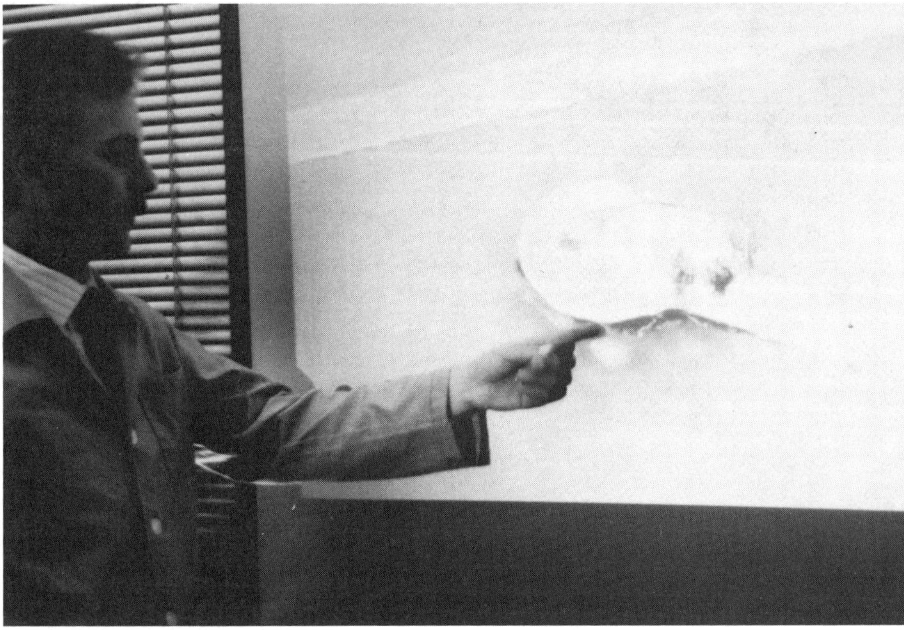

Figure 6.15 Trainer looking at a solder joint. (Courtesy of OMNI Training Corp., Cucamonga, CA.)

retrains the operators when they are released into manufacturing. This must be monitored continuously to certified compliance to the program. Engineer training must come from attending seminars with continuous education programs so that engineers may keep up with the latest technological advances. Quality personnel are the representatives of your customers and they must be fully aware of the customer requirements and the need for product reliability and customer satisfaction.

6.3 THE THEORY OF MANUAL SOLDERING

6.3.1 Theoretical Approaches

The theoretical approaches to manual soldering must include all the elements previously discussed. It is meaningful that the operators understand the mechanics of creating a solder joint. This knowledge should include the total explanation of thermal profiling which occurs during the hand-soldering process. Thermal profiling is the actual temperature differentials which are created when the solder joint is made. It identifies when the soldering iron is idling and when it is being used and how much temperature loses are incurred when the solder joints are made. Without this knowledge, improper skills will be developed potentially creating more problems than they are trying to solve.

Carrying heat to the work is the concept of manual soldering. The temperature of the iron tip is an ambiguous number which must be properly understood to design the manual soldering process.

Tip temperature implies the following:

- Actual tip temperature
- Caloric content of the tip
- Recovery cycle
- Wattage of the iron

Hence, all manual soldering functions performed must consider these four elements. If any are improperly specified the correct size iron and tip will not be chosen and the solder quality will be affected.

Tip temperature is self-explanatory; however, caloric content of the tip, implies the amount of heat the tip can store while being used and recovery cycle is the amount of time it takes the soldering iron to reach working temperature immediately after creating the preceding solder joint. Wattage is the energy input to the solder iron to maintain the proper energy levels needed to maintain the tip temperature. Figure 6.16 identifies the working zone of both thermal control and standard type of solder iron. As can be seen, better temperature controls are evident with thermal control irons. Working tip temperature should stay as close to idle temperature as possible during cyclical soldering. This means the tem-

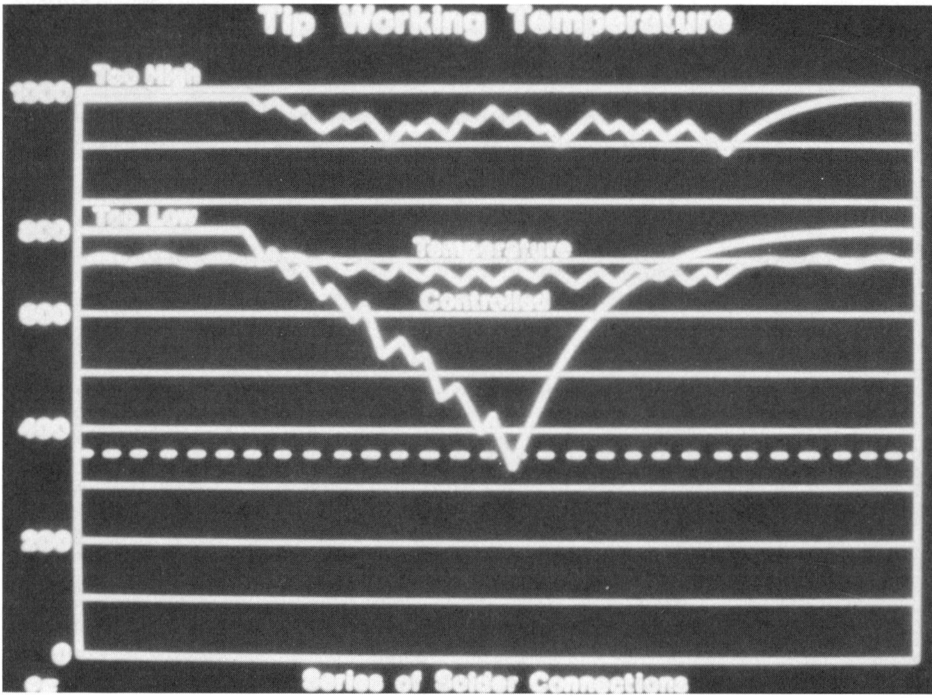

Figure 6.16 Graph of thermal profiie. (Courtesy of OMNI Training Corp., Cucamonga, CA.)

perature drop which occurs when a solder joint is made due to the heating sinking effect of the solder joint should be kept to a minimum. Then, when the solder joint is completed and the solder iron removed, the wattage energy of the iron makes up the temperature loss, and this is called recovery time of the soldering iron.

6.4 SOLDERING WORKSTATION

Having discussed temperature cycling of the iron, let's review the entire manual soldering workstation. First and foremost is the organization of the workbench, i.e., the tools and materials to do the job, in the placement of the work (see Fig. 6.17).

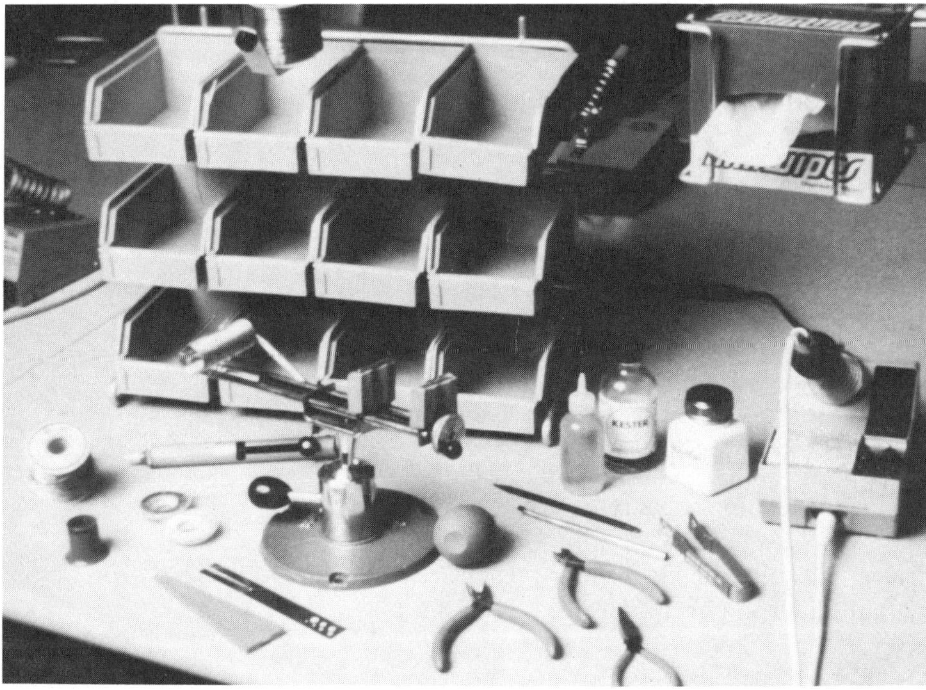

Figure 6.17 Workbench layout. (Courtesy of OMNI Training Corp., Cucamonga, CA.)

Soldering workstations should be organized for easy access to all the tools. Typical tools must include:

- Pliers, bent-nose and straight
- Wire strippers for each wire size
- Soldering aid
- Tweezers
- Wire cutters
- Lead-forming tools
- Wet sponge

- ○ Solder iron holder (could be more than one if demanded by a variety of joint sizes)
- ○ Solder iron
- ○ Spare solder iron tips
- ○ Screwdrivers
- ○ Tool box
- ○ Board holder
- ○ Cored solder
- ○ Flux
- ○ Venting fan (option)
- ○ Cleaning solvent

Keep all solder irons in their respective holders. Do not put more than one iron per holder.

6.5 SOLDERING TECHNIQUES

6.5.1 Wire Preparation

Preparing the wires for soldering requires the stripping of the insulation from either the core strand wire or the twisted bundle. The insulation-stripping operation must not nick the wire or cut any of the strands (Table 6.1). Stripping the wire is performed with various types of tools (mechanical, automatic, or heat-melt systems). Figure 6.18 depicts an automatic wire-stripping system, while thermal strippers are shown in Fig. 6.19 and mechanical strippers are shown in Figs. 6.20 and 6.21. Although these will not be discussed in this text, it must be understood that no damage to the wire is the preferred condition (see Fig. 6.22) for properly stripped wire, while minimal damage such as minute nicks or broken strands is at times acceptable as the minimal acceptable condition. Nicks and broken strands cause a reduction in the total cross-sectional area of the wire, which must be

Table 6.1 Limits of Stranded Wire Damage

Number of strands	Maximum allowable nicked or broken strands
Less than 7	0
7-15	1
16-18	2
19-25	3
26-36	4
37-40	5
41 or more	6

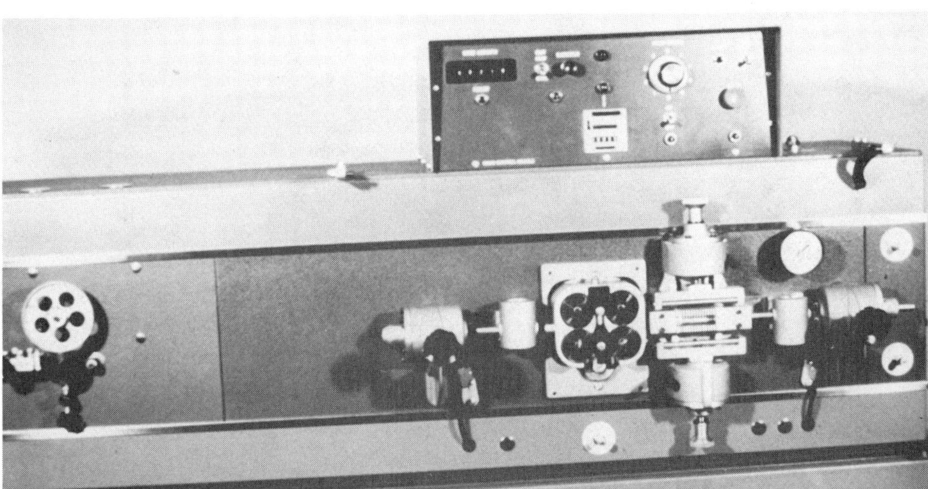

Figure 6.18 Autowire stripper. (Courtesy of OMNI Training Corp., Cucamonga, CA.)

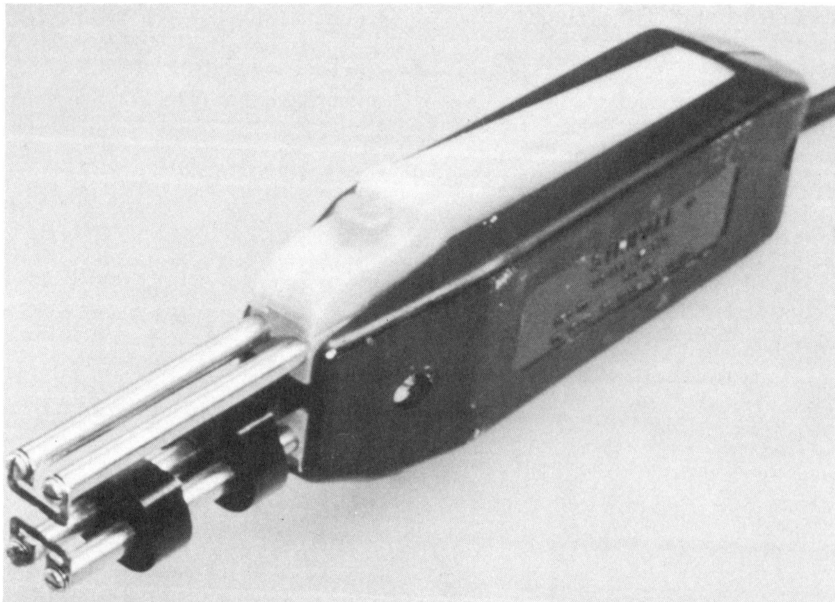

Figure 6.19 Thermal strippers. (Courtesy of OMNI Training Corp., Cucamonga, CA.)

compensated for in the electrical design. The basic standards used by industry do allow for some cut strands on the stranded wire but do not allow nicks in small-gage solid wire.

Strip length of insulation is critical in certain applications to prevent shorting to adjacent wires or components. The typical strip length as shown in Fig. 6.23 is 1/8 in. maximum, and the minimum is governed by the rule, "Don't get the insulation into the solder joint." This is critical to prevent insulation damage as shown in Fig. 6.24 and solder wicking beneath the insulation. Such wicking is not allowed because it reduces the stress relief capabilities of the wire and prevents cleaning of the residual fluxes that have wicked up beneath the insulation. Operator training is a must in order to develop the proper skills required for manual soldering of wires.

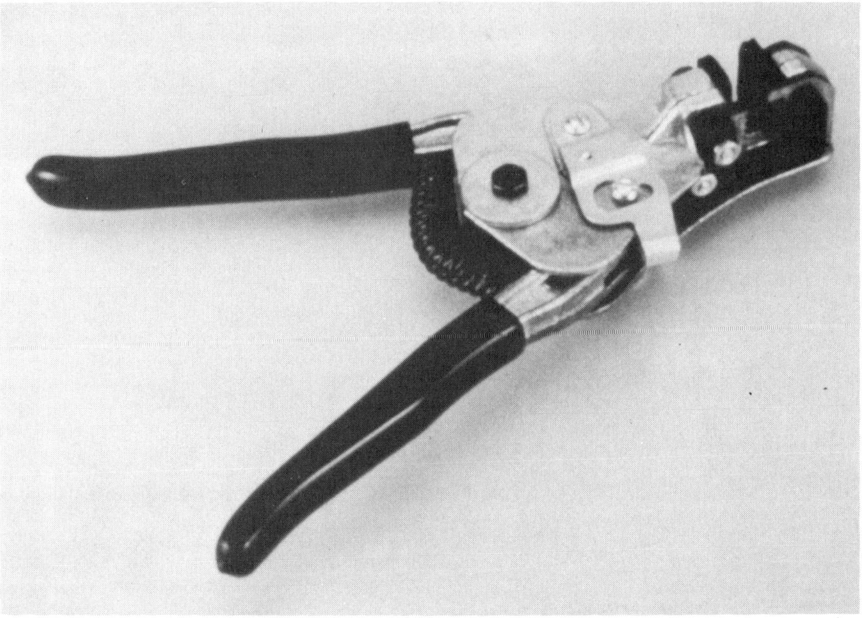

Figure 6.20 Mechanical strippers. (Courtesy of OMNI Training Corp., Cuamonga, CA.)

6.5.2 Pretinning and Mechanical Attachment

Once stripped, the stranded wires should be pretinned as shown in Fig. 6.25, to prevent bird caging and to ease the securing of the wire to any termination. When creating a solder joint with a wire to a terminal, a good electromechanical connection is necessary. All wires must be mechanically secured to the terminal prior to the soldering operation (see Fig. 6.26). This is accomplished by wrapping the wire around the terminal post. The wrap must be at least 180° around the terminal post and no more than 360° (see Fig. 6.27). With the wire wrapped around the post securely prior to the soldering function, it won't move during the soldering operation, leaving both hands free to bring the solder and solder iron to the joint.

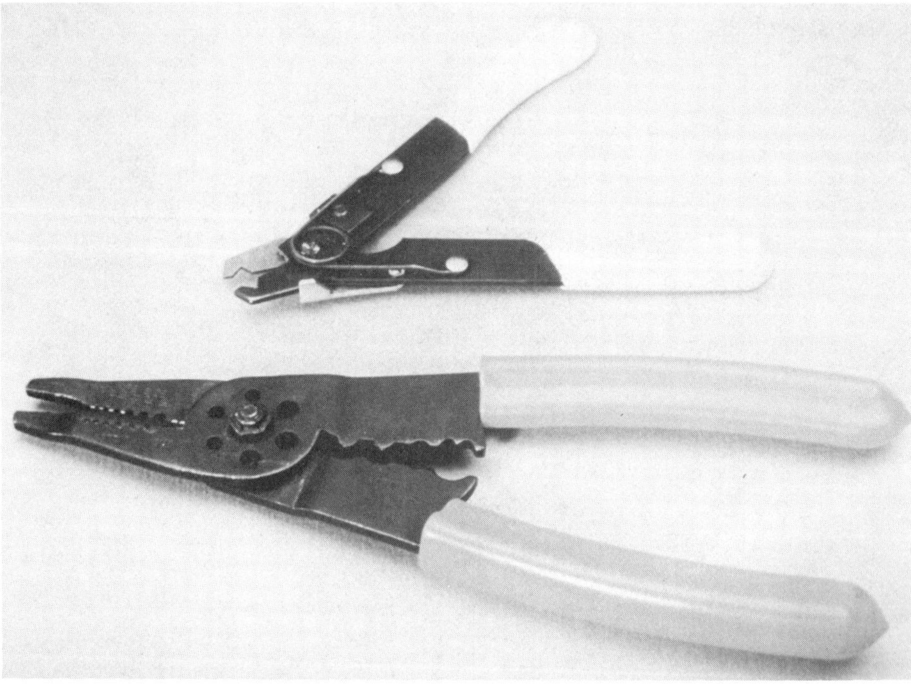

Figure 6.21 Two mechanical strippers. (Courtesy of OMNI Training Corp., Cucamonga, CA.)

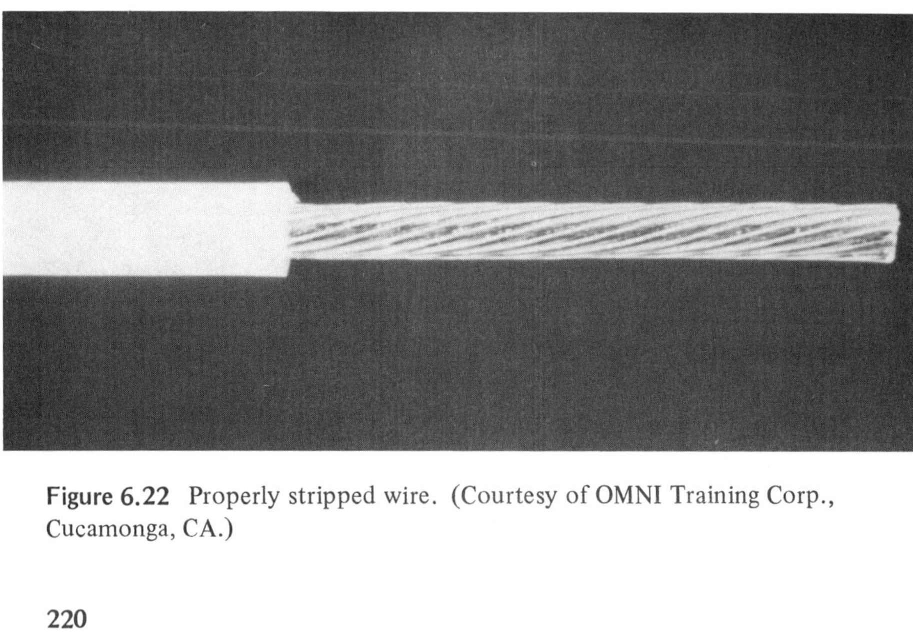

Figure 6.22 Properly stripped wire. (Courtesy of OMNI Training Corp., Cucamonga, CA.)

Figure 6.23 Terminal Post. Wire wrap good gap. (Courtesy of OMNI Training Corp., Cucamonga, CA.)

It must be noted that the direction in which the wire is wrapped around the post is critical. The wrap must be made so any forces applied to the wire will not create peel forces against the solder joint (see Fig. 6.28). Any forces applied to the wire should be absorbed by both the solder joint and the terminal post.

6.5.3 Terminal Post Soldering

The proper method to make the solder joint once the wire is mechanically secured is as follows:

1. Wipe iron on wet sponge to remove oxidized material from the tip of the tool, as shown in Fig. 6.29.

Figure 6.24 Terminal Post, wire wrap, insufficient insulation gap. (Courtesy of OMNI Training Corp., Cucamonga, CA.)

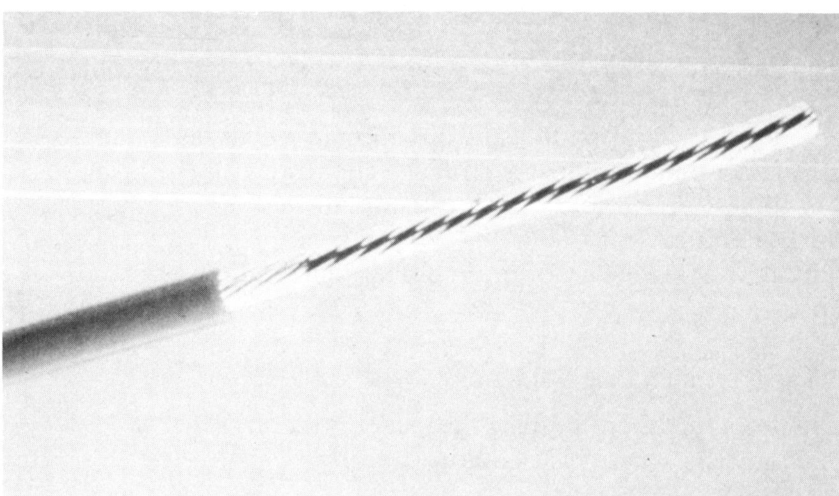

Figure 6.25 Tinned Wire. (Courtesy of OMNI Training Corp., Cucamonga, CA.)

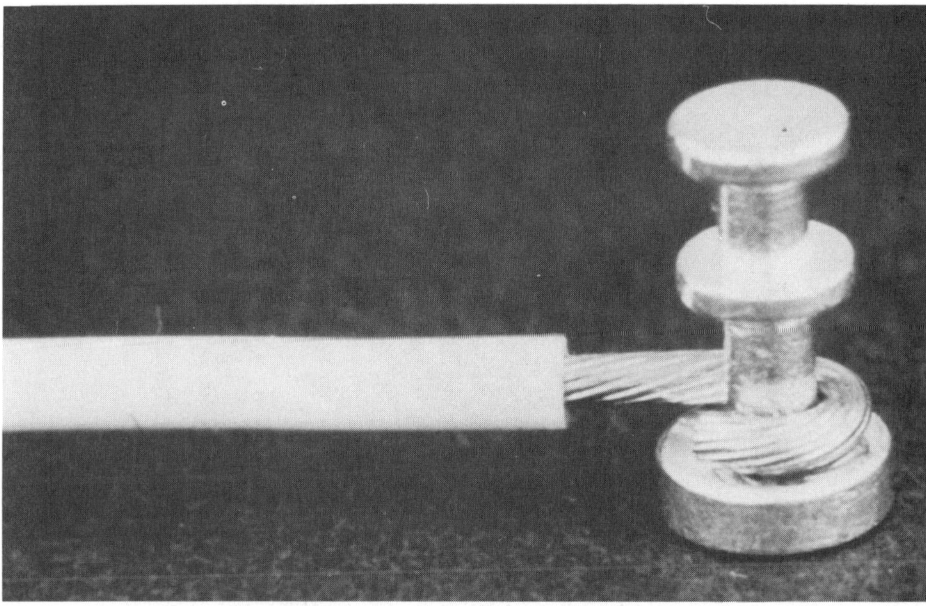

Figure 6.26 Terminal post, wire secured. (Courtesy of OMNI Training Corp., Cucamonga, CA.)

2. Using cored solder, apply a small amount onto the tip of the solder iron, as shown in Fig. 6.30.
3. Place the iron onto the opposite side from which the solder will be brought into the joint, as shown in Fig. 6.31. Make sure the largest surface of the tip is in contact with the joint. This will enhance the transfer of heat from the solder iron tip to the joint.
4. Apply the solder from the opposite side of the solder iron tip, as shown in Fig. 6.32. Once the core solder starts to flow, allow it to flow completely around the terminal post and the wire being joined.
5. Make sure the following exists:

    ○ No damage to the insulation

Figure 6.27 Terminal post, wire wrap. (Courtesy of OMNI Training Corp., Cucamonga, CA.)

    - No charring
    - No burning
    - No melting
    - No shrinking
- No excessive solder
    - If solid wire is used, the outline of the wire must be visible, as shown in Fig. 6.33.
    - If stranded wire is used, the outline of the outer strands must be visible and a wetting flow should be visible to the post.

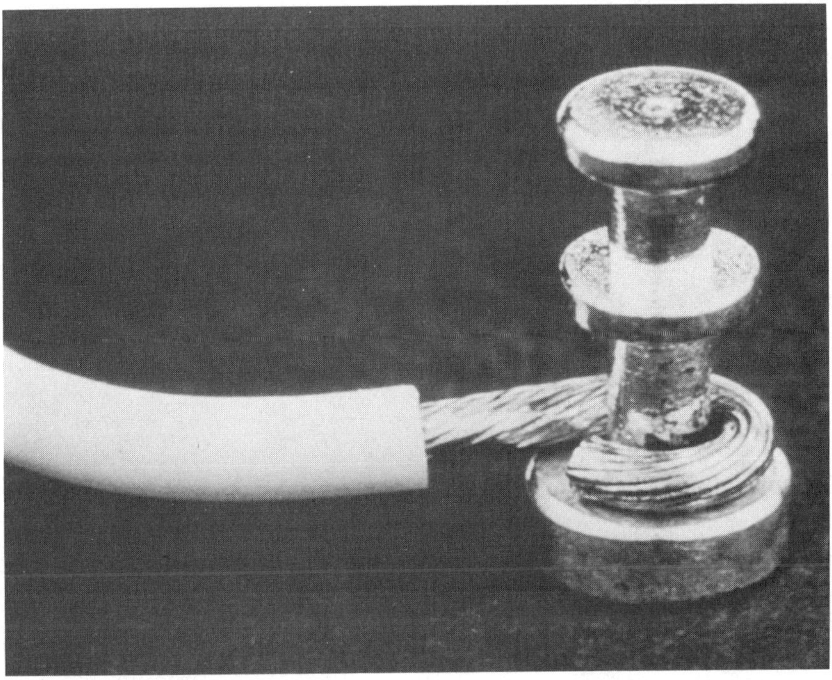

Figure 6.28 Terminal post, improper wire direction. (Courtesy of OMNI Training Corp., Cucamonga, CA.)

6. Cleaning the solder joint of residual fluxes as viewed in Fig. 6.34.

One must review the final requirements of the product and the flux materials used in the cored solder. Many mildly activated fluxes (RMA) do not need to be removed, while all rosin activated (RA) should be removed.

Water-soluble flux cored solder is not recommended for soldering any types of wire, stranded or solid. This is because the activity levels of the flux under any and all conditions is corrosive and the residues must be removed. When soldering any wire, the flux has a tendency to wick up the wire beneath the insulation. Once this occurs, the flux can no longer be removed and degrada-

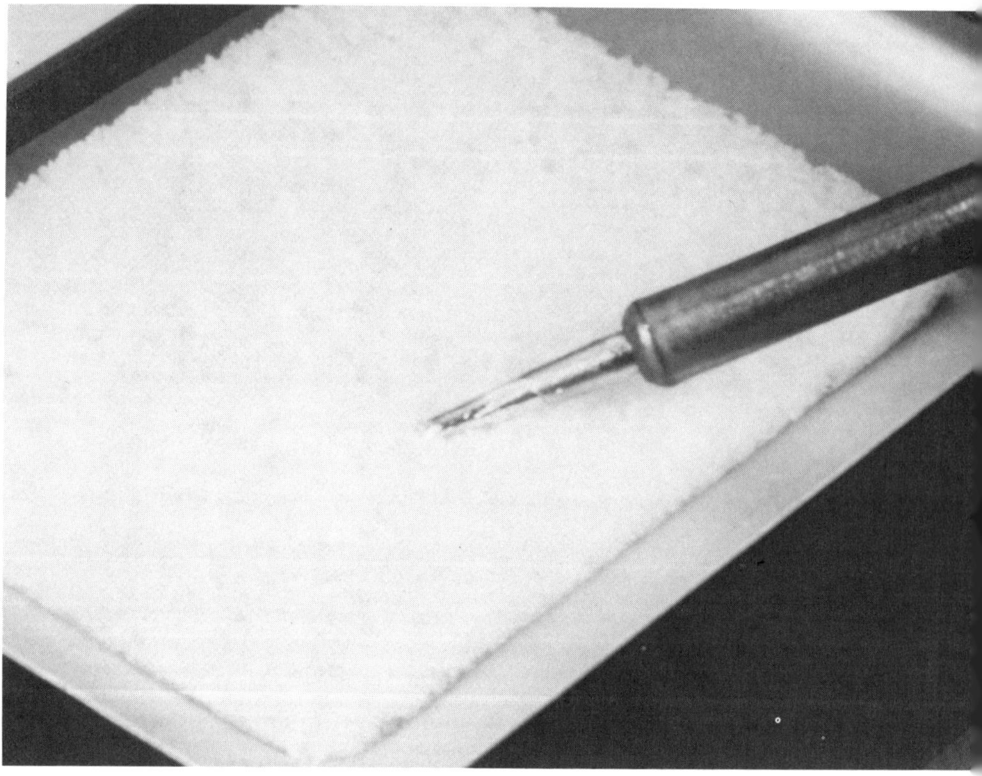

Figure 6.29 Solder iron on sponge. (Courtesy of OMNI Training Corp., Cucamonga, CA.)

tion and corrosion of the wire will start. This is an unwanted condition and must be avoided under all conditions.

6.5.4 Manual Soldering of Components into Printed Wiring Assemblies

Soldering components into plated through holes requires a knowledge of the board construction and materials including component lead materials. For the solder to properly flow up into the plated through holes, the materials (i.e., the barrel and the leads) must reach soldering temperatures. If they don't reach the melt-

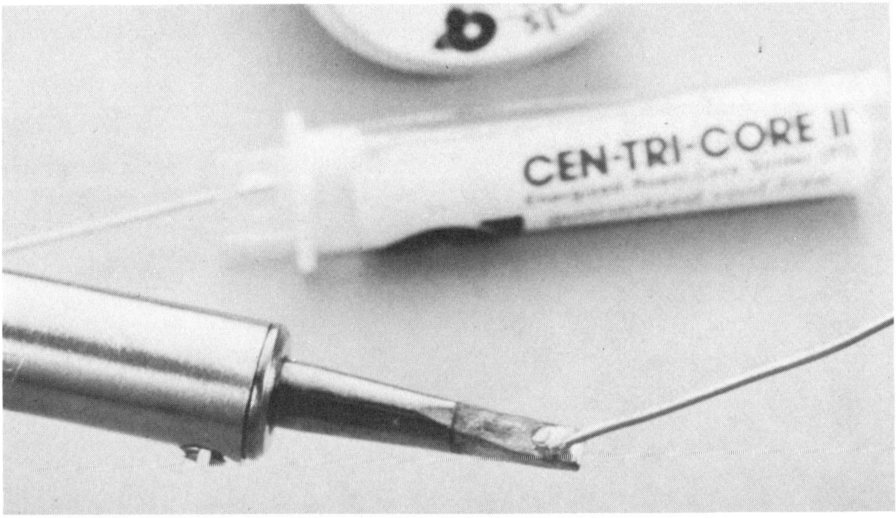

Figure 6.30 Tinning of solder iron tip. (Courtesy of OMNI Training Corp., Cucamonga, CA.)

Figure 6.31 Terminal post, solder iron on terminal post. (Courtesy of OMNI Training Corp., Cucamonga, CA.)

Figure 6.32 Terminal post, solder being added to create solder joint. (Courtesy of OMNI Training Corp., Cucamonga, CA.)

ing temperature of the solder, the molten solder coming off the solder iron will solidify and the filling of the plated through hole will not occur.

The filling of the plated through hole with solder is directly related to the heat transfer up into the plated through hole. As the temperature of the barrel's metal rises above the melting temperature of the solder, the solder wicks up into the barrel filling it completely (see Fig. 6.35). Additional metal such as a component lead enhances the wicking of solder because of its physical size and the heat it captures when contacted by the solder iron. This additional heat increases the caloric or heat content of the plated through hole allowing the solder to remain molten and rise up into the plated through hole. Conversely, any metallic inner layers, such as ground and voltage layers, act as heat sinks to remove heat from the plated through hole, as shown in Fig. 6.36. This heat re-

Figure 6.33 Resistor terminated to terminal post. (Courtesy of OMNI Training Corp., Cucamonga, CA.)

moval prevents the barrel from heating to the required temperatures, preventing the solder from wicking up completely through the plated through hole resulting in a partially filled hole. Therefore to get the solder to rise up into the plated through hole, the concept is to apply enough heat to the joint to raise the temperature above the melting point of solder.

6.5.5 Touchup Soldering

Another area of soldering that requires maximum skills is touchup soldering of printed wiring boards. The term touchup is probably the most misused term in the industry. Originally intended to remove resultant icicles and shorts on assemblies soldered by the wave-soldering process, it is now used to repair damaged or defec-

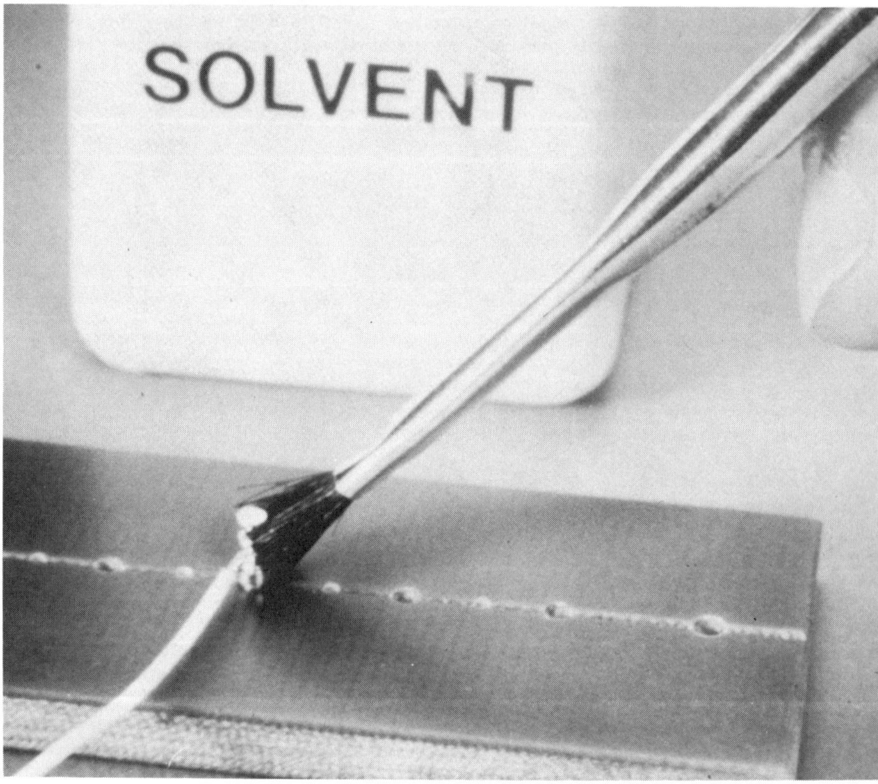

Figure 6.34 Cleaning of terminal post. (Courtesy of OMNI Training Corp., Cucamonga, CA.)

tive solder joints. Reemphasizing, it is now the term used to bring an assembly up to its final quality, including soldering components into place that were not soldered during the mass reflow process. The term today is inclusive of the following repairs:

- Defective joints
  - Disturbed–Fig. 6.37
  - Cold solder–Fig. 6.38
  - Grainy–Fig. 6.39

Figure 6.35 Resistor, good solder flow through on component side. (Courtesy of OMNI Training Corp., Cucamonga, CA.)

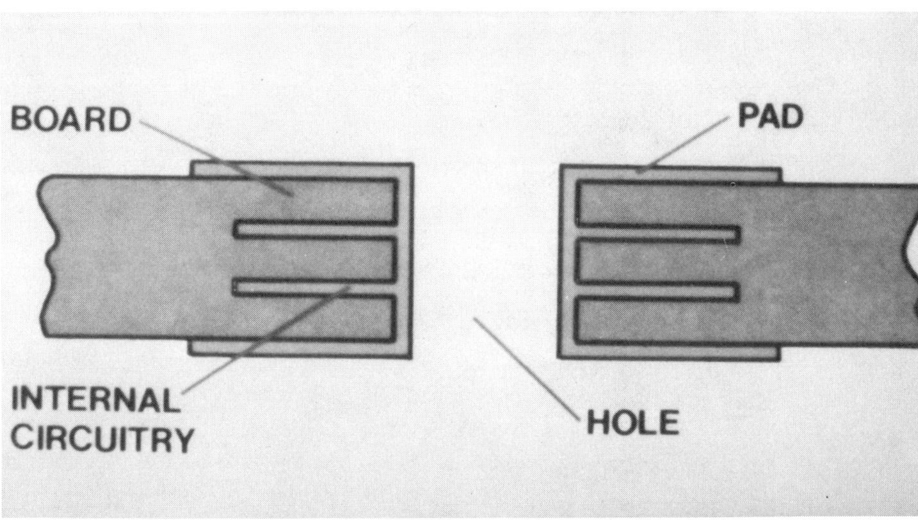

Figure 6.36 Cross-sectional sketch of plated through hole, multilayer board. (Courtesy of OMNI Training Corp., Cucamonga, CA.)

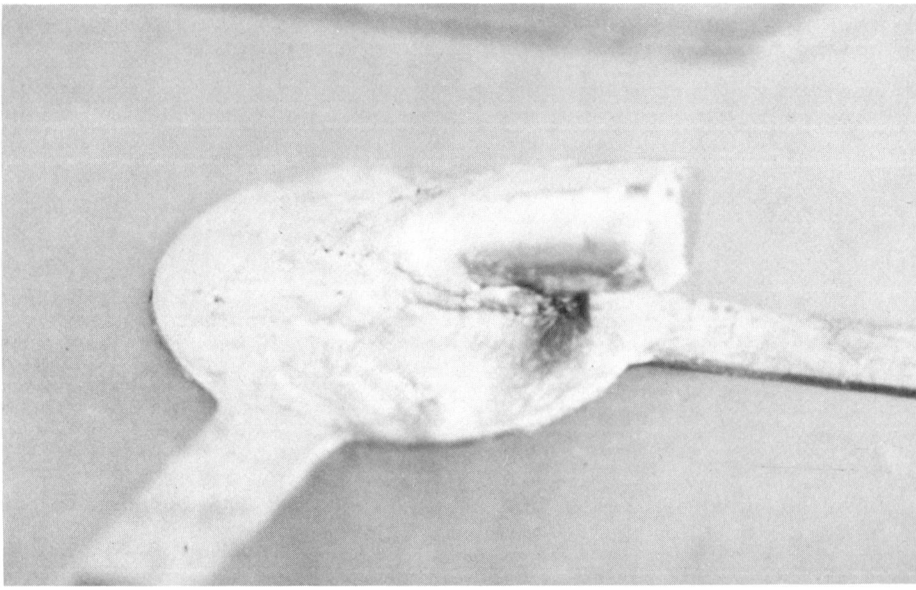

Figure 6.37 Disturbed solder joint. (Courtesy of OMNI Training Corp., Cucamonga, CA.)

  - ○ Nonshiny
  - ○ Lifted pads–Fig. 6.40
- ○ Shorts
  - ○ Lead to lead–Fig. 6.41
  - ○ Lead to etch–Fig. 6.42
  - ○ Etch to etch–Fig. 6.43
- ○ Nonwetting–Fig. 6.44
- ○ Dewetting–Fig. 6.45
- ○ Missing component–Fig. 6.46
- ○ Measles–Fig. 6.47

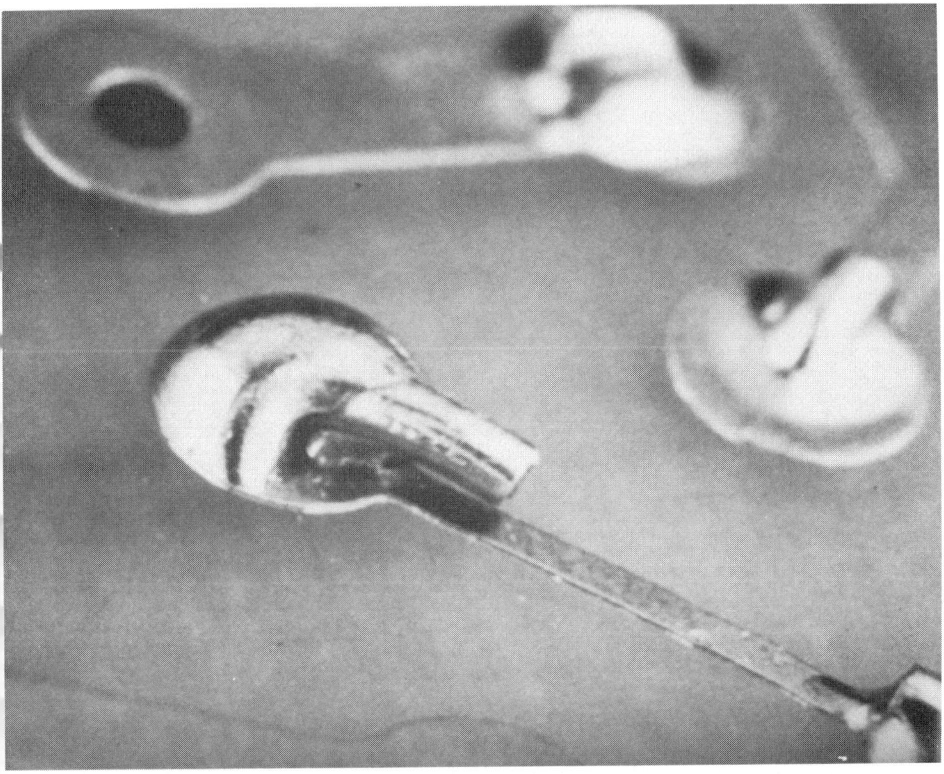

Figure 6.38 Cold solder joint. (Courtesy of OMNI Training Corp., Cucamonga, CA.)

Touchup Cost

Generally touchup soldering is the most extravagant wasteland of manual labor and its cost must be constantly reviewed for economic justification by all manufacturing, quality, and test engineers. Data collection is a must and all defects must be charted to define the priority of attack. All defects must be identified by skilled operators so that touchup is performed only when necessary. Touchup operators left to themselves will touch up every joint on an assembly, which is a needless, wasteful use of resources, and it impairs product reliability.

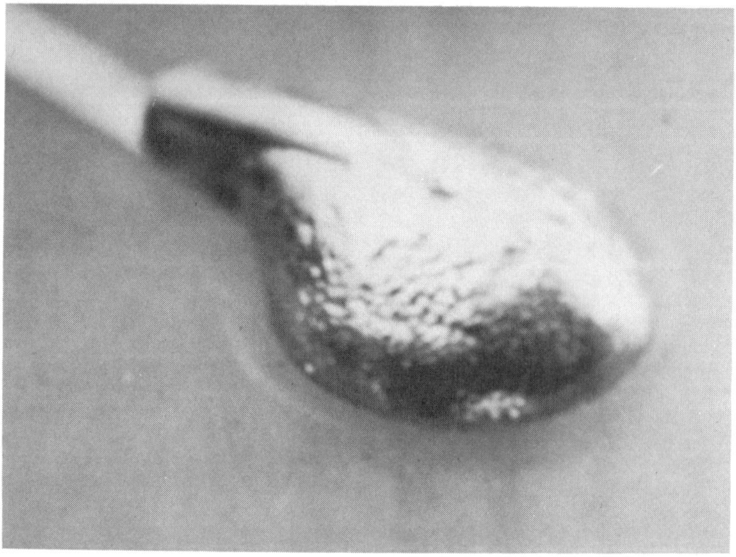

Figure 6.39 Grainy solder joint. (Courtesy of OMNI Training Corp., Cucamonga, CA.)

To prevent needless touchup, the operation must be closely monitored. In theory, the touchup function is to repair functional defects; however, it is misused, and consequently many solder joints are reflowed simply to improve cosmetic appearance. Many studies performed identify that touched up solder joints fail quicker in reliability testing. This is an important point to remember: touchup solder joints are more susceptible to failure than ones that are not touched up.

Therefore to prevent excessive touchup, a two-step touchup operation is strongly recommended. Ordinarily this operation consists of an inspector who identifies and marks the defects per the product requirements. The defect is then reparied per the product repair procedures, as specified per corporate specifications. The engineer defining the touchup operation must be aware of cosmetic rejects and determine whether or not it is economical to alter the visual appearance.

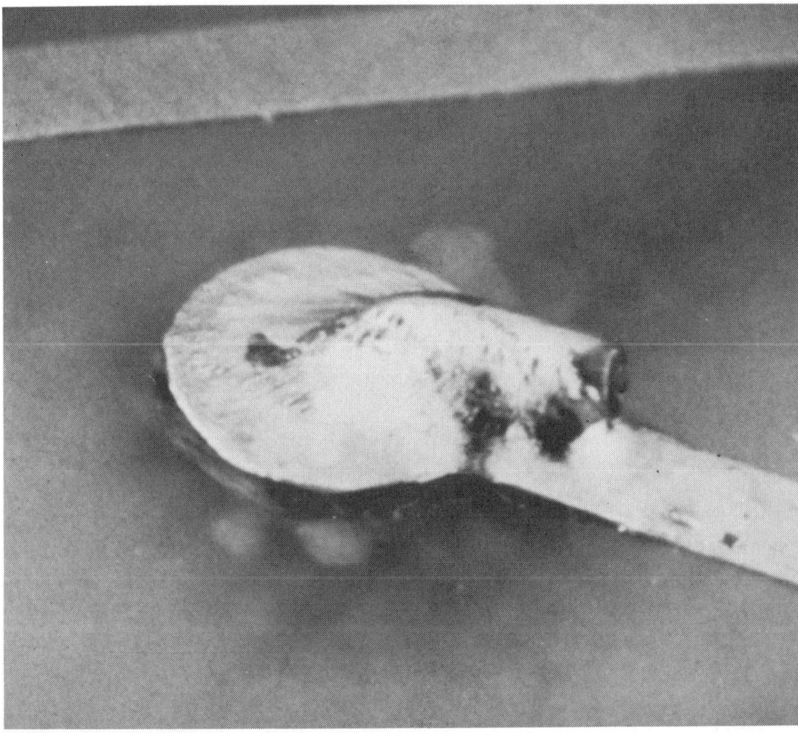

Figure 6.40 Lifted pad. (Courtesy of OMNI Training Corp., Cucamonga, CA.)

Controlling Touchup

A major problem in industry is the lack of control in identifying total defects per joints created and defining the financial ramifications of such rework. Identifying these elements are necessary to control the process and produce quality product. Hence all engineers should undertake the following challenges:

1. When designing manufacturing facilities, do not include touchup areas; send all product directly to test.

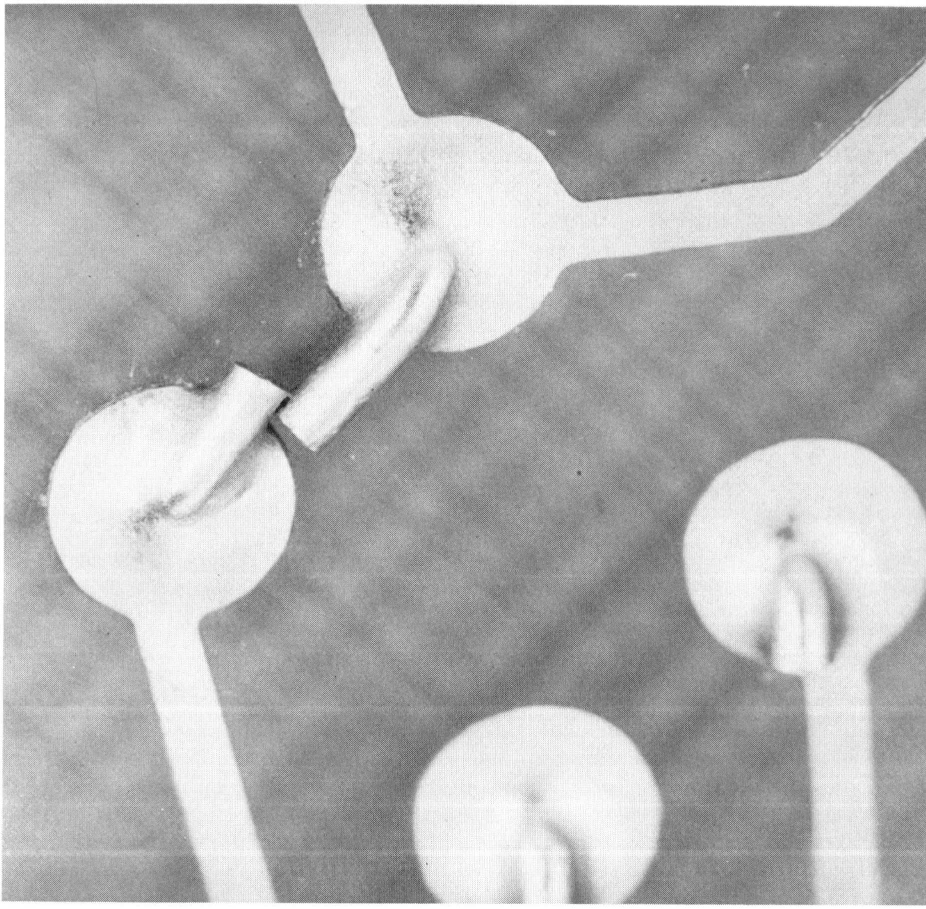

Figure 6.41 Lead-to-lead short. (Courtesy of OMNI Training Corp., Cucamonga, CA.)

2. In touchup areas that do exist, remove only solder shorts and touch up the missing solders and move that product into electrical test.
3. Reduce work in process.

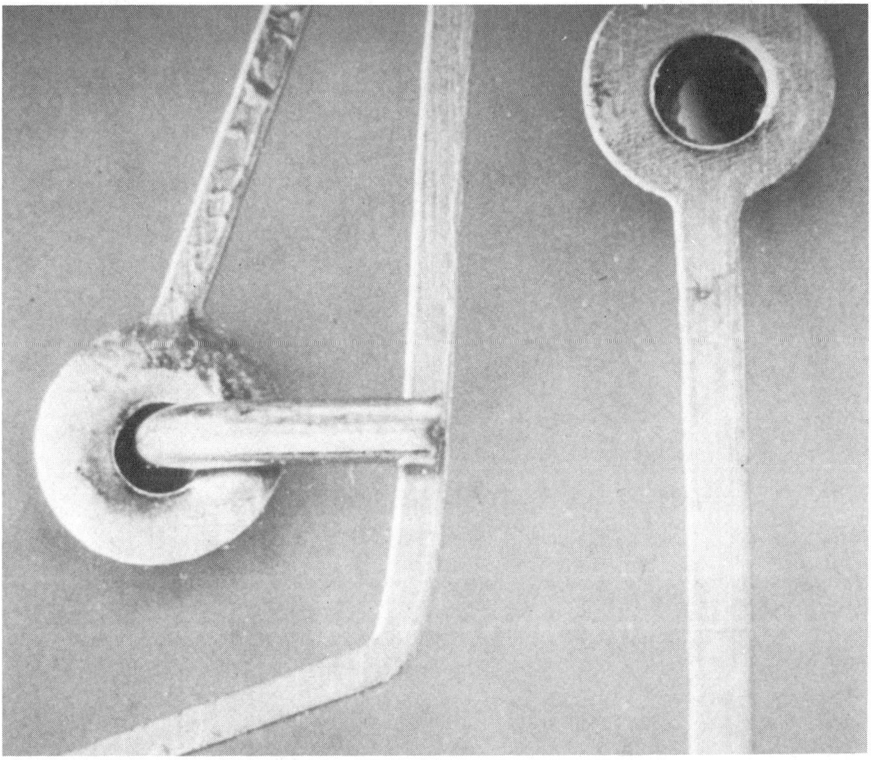

Figure 6.42 Lead-to-etch short. (Courtesy of OMNI Traing Corp., Cucamonga, CA.)

As simplistic as this may sound, it is the most difficult challenge now offered to manufacturing engineers responsible for board assembly processes.

6.5.6 Repair Process

Another domain of manual soldering in the manufacturing of printed wiring is the component removal and replacement process. This operation is performed in conjunction with electrical test and repairing the defects found from those particular tests.

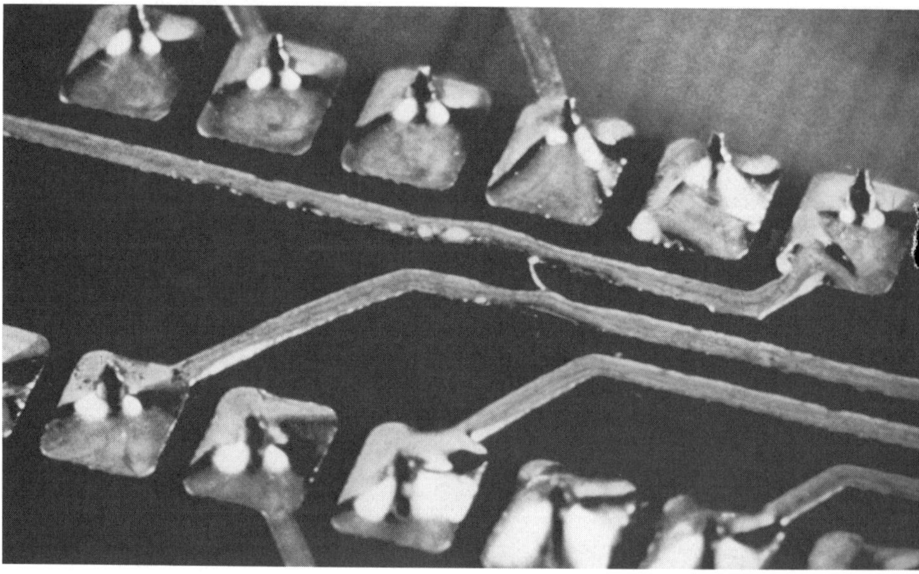

Figure 6.43 Etch-to-etch short. (Courtesy of OMNI Training Corp., Cucamonga, CA.)

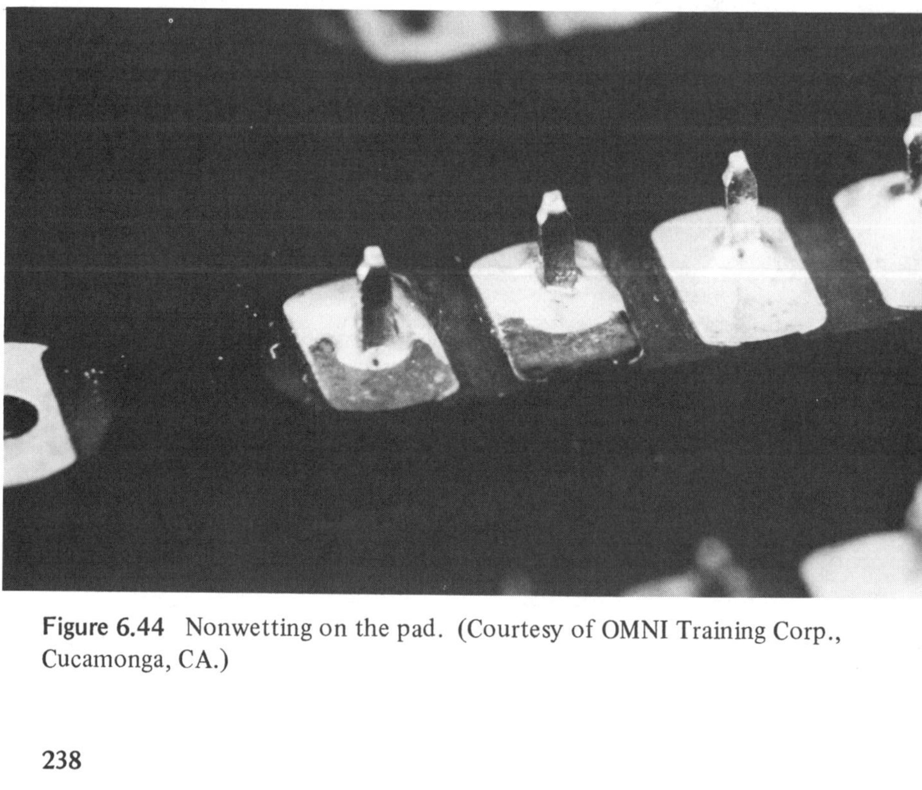

Figure 6.44 Nonwetting on the pad. (Courtesy of OMNI Training Corp., Cucamonga, CA.)

Figure 6.45 Dewetting on strip area. (Courtesy of OMNI Training Corp., Cucamonga, CA.)

This process is very exacting and must be performed by qualified personnel. Qualified personnel implies that the operators have had the proper training and are certified to do the job. Equipment selection and operator training are the major elements necessary to prevent any damage and to produce a quality product. Without proper training and forethought by the engineering staff, this operation will add dollars to the cost of the product. It must be remembered that this is not value added to the product, but an overhead expense, and therefore it should be thought of as an added cost that impairs the reliability of the product. As can be seen, the theme of assembly must be to do it right the first time! To investigate equipment and operator training, individual attention must be given to the following:

Figure 6.46 Missing component. (Courtesy of OMNI Training Corp., Cucamonga, CA.)

- Component removal process
  - Tool category selection
    - Manual and semiautomatic
- Component replacement process
  - Operator selection
  - Operator training
  - Workplace layout
  - Quality requirements

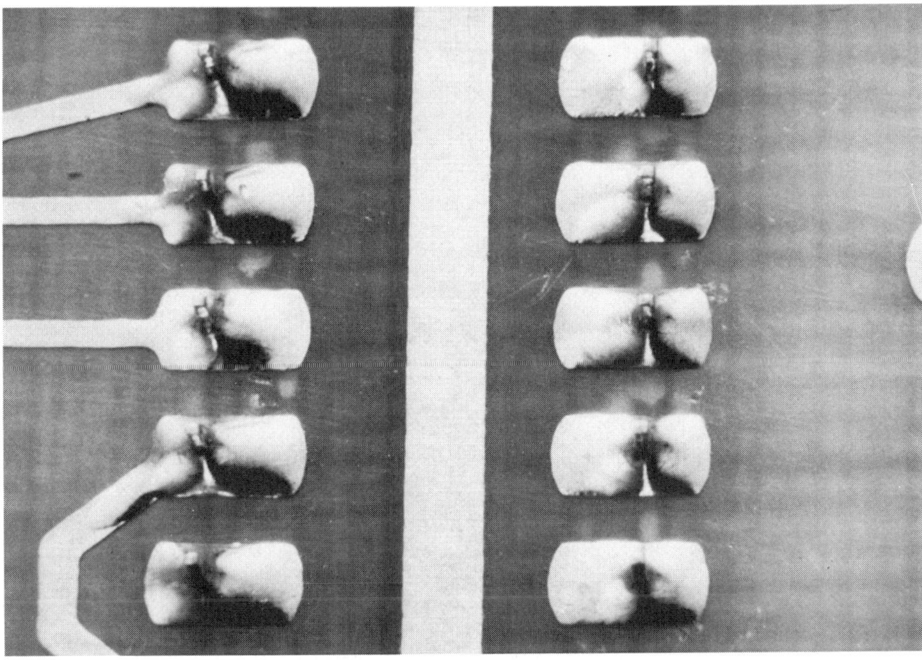

Figure 6.47 Measling, whitish spots around pads. (Courtesy of OMNI Training Corp., Cucamonga, CA.)

Component Removal

Reasons for component removal are:

- Electrical failure of components
- Board defects beneath components
- Test parameters not sophisticated enough to identify between good and bad components necessitating the removal of multiple components
- Engineering change orders

The component removal techniques are dependent on the variety of component configurations and styles. Axial leaded

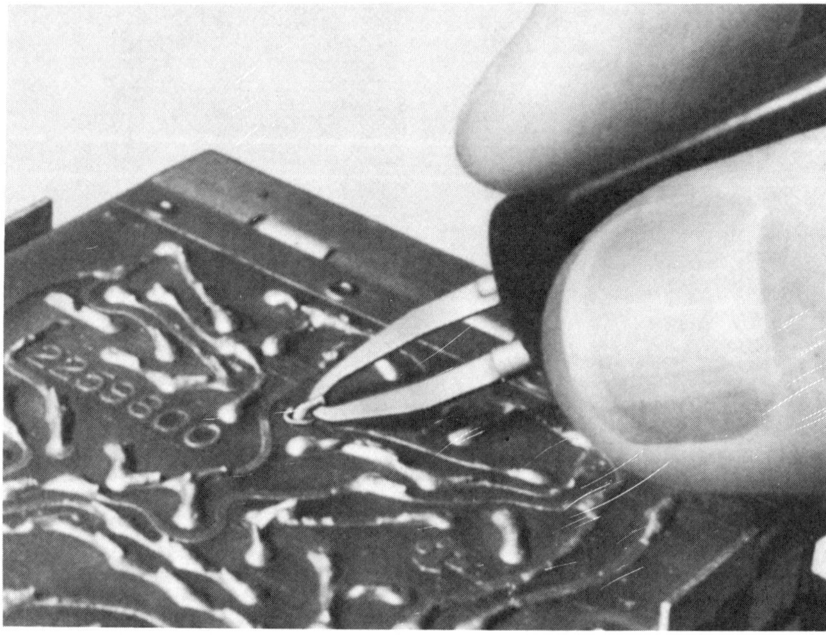

Figure 6.48 Use of thermal parting tool to melt solder and then lift the lead. (Courtesy of Pace Incorporated, Laurel, MD.)

components, i.e., resistors, capacitors, LEDs, etc., are usually destroyed when removed due to their low cost. Their leads are mechanically cut, as shown in Fig. 6.48, leaving a portion of the lead in the soldered plated through hole to be removed at a secondary operational step. In removing the lead the solder is reflowed, the lead is removed, then the solder is evacuated from the plated through hole with a solder removal tool. This is performed for ease of installing the replacement part.

For the components removal process, define the components that must be removed and what is to happen to that component once it is removed. Will it be scrapped? Will it be retested and reused? The latter could be the case if the problem failure was related to the circuitry rather than the particular component.

There are many methods for removing components from printed wiring boards. Whichever method is used, the solder within the plated through holes must either be removed to free up the lead or all the plated through holes must be simultaneously reflowed and kept molten until the component is removed.

Knowledge of the soldering operation will identify the differences between initial soldering and component removal and replacement. When a solder joint is initially made, the heat is applied to both the board and component lead and the heat energy is absorbed by the surrounding metallization and used to reflow the cored solder material. While repairing or removing a component, the reflow operation is accomplished by placing the iron on the solder joint in the same fashion as the initial solder joint. However, the heat transfer through the board is different. The main difference is the mass of metal which must be heated and reflowed. The volume of solder in a filled plated through hole is 75-80% higher than in an empty hole. This requires more energy for complete reflow. Energy is measured as temperature through some time period; therefore, identifying the solder iron as constant, the variable is the time necessary to bring the joint to a molten condition. This extra time must be defined to reflow the solder joint. Reflowing the solder joint implies that the entire mass is brought to molten condition. Therefore to properly desolder a solder joint, define the correct solder iron, tip size and configuration, and procedural methods. If this step is not performed correctly, many boards will be needlessly scrapped.

Various tools and equipment are available to perform these tasks. However, certain precautions must be taken in the training programs to develop the skills in the usage of the tool to preventing damages that can be experienced with unskilled operators.

Repair Equipment Selection

Equipment selection must also be critically reviewed so that the chosen tools do not damage the components or the assembly. The determining factor in choosing equipment is the ability of the tool to reflow the solder in the plated through hole within as short a period of time as possible. The tool selection process must be developed through an understanding of what has to be repaired or removed, i.e., component types, laminate materials, and the volume of repair. With this information identified, the equipment selection process is simplified so that one can stay within the guidelines of the experiment. In either case, financial justification is a must to properly justify any equipment.

The second step in choosing the equipment is to perform the required tasks and developing experiments to prove the feasibility of the equipment. Some equipment has been designed to remove more than one type or style of component, while others are designed to remove only solder from the plated through holes. In any case, the major function of any repair operation is to reflow the solder in the plated through hole, remove such solder, or remove the component from the molten solder.

6.5.7 Removal Methods

Methods of component removal are:

1. Heat and shake, (not recommended)
2. Heated vacuum
3. Heat and vacuum draw
4. Heated braid
5. Hot air reflow
6. Preconfigured solder flow wells

Heat and Shake

This method is identified for historical purposes only and not recommended for any manufacturing or repair operation. The process consists of heating the solder in the plated through hole to a molten condition, then grasping the assembly and giving it a sharp blow against a solid object such as a workbench. The quick deceleration of the assembly is not imparted to the molten metal and when the board stops against an object, the molten solder continues and exits the hole in a slug type of fashion. The shock on the printed wiring assembly is detrimental to its functionality and reliability.

Vacuum Solder Removal

There are two types of vacuum removal tools:

1. Heated–Fig. 6.49
2. Ambient–Fig. 6.50

How does one choose or decide which unit is best for any particular application; Consider the following when making the choice:

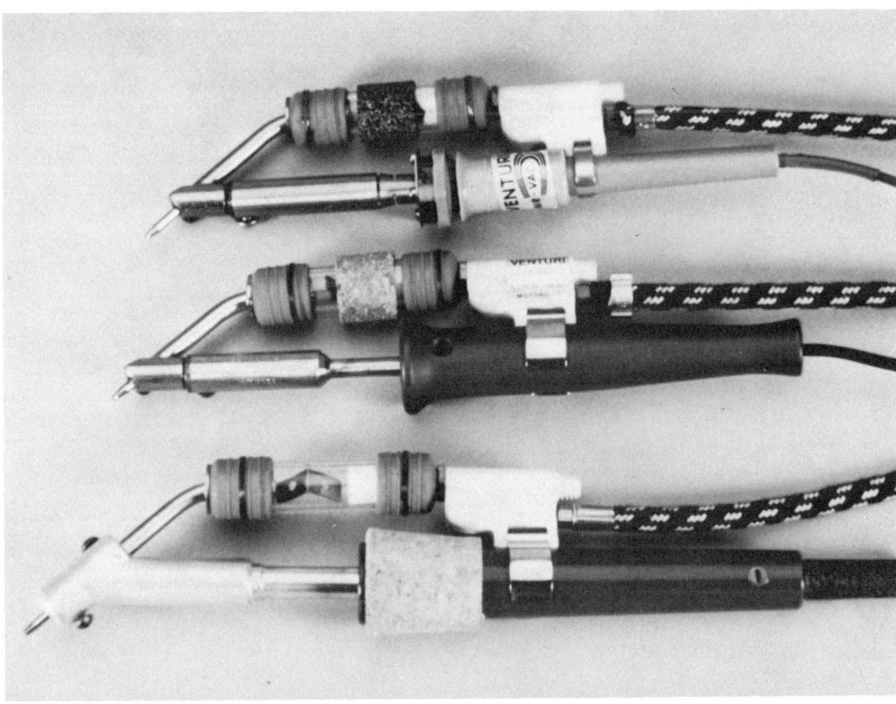

Figure 6.49 Heated vacuum tool. (Courtesy of OMNI Training Corp., Cucamonga, CA.)

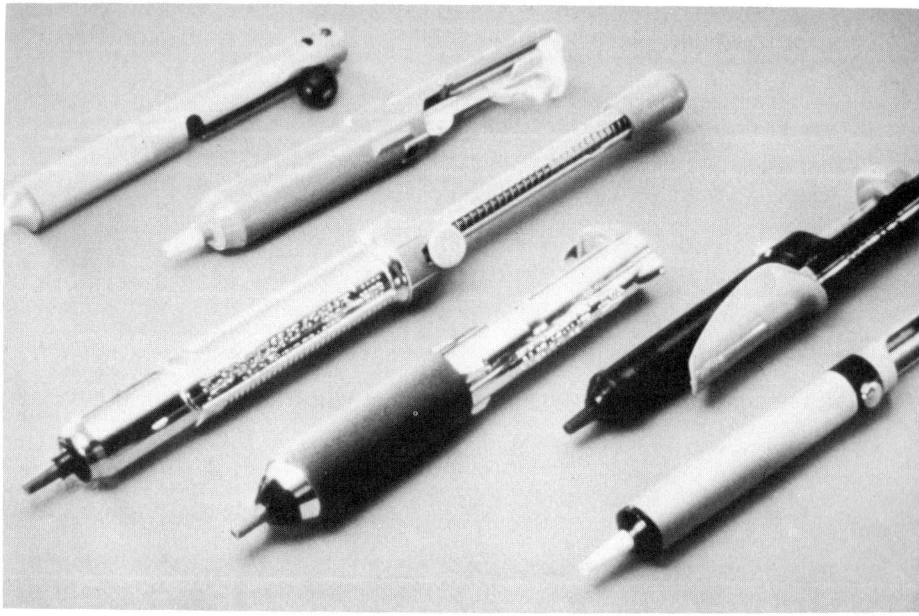

Figure 6.50 Ambient Temperature vacuum tool. (Courtesy of OMNI Training Corp., Cucamonga, CA.)

- ○ Heat source and power
- ○ Vacuum source
    - ○ Self-contained
    - ○ Facility supplied
- ○ Weight and design of tool
- ○ Workbench layout
- ○ Tip size and material—Fig. 6.51
- ○ Maintenance

The first phase is to define the components to be removed and then determine the best way to remove them without destroy-

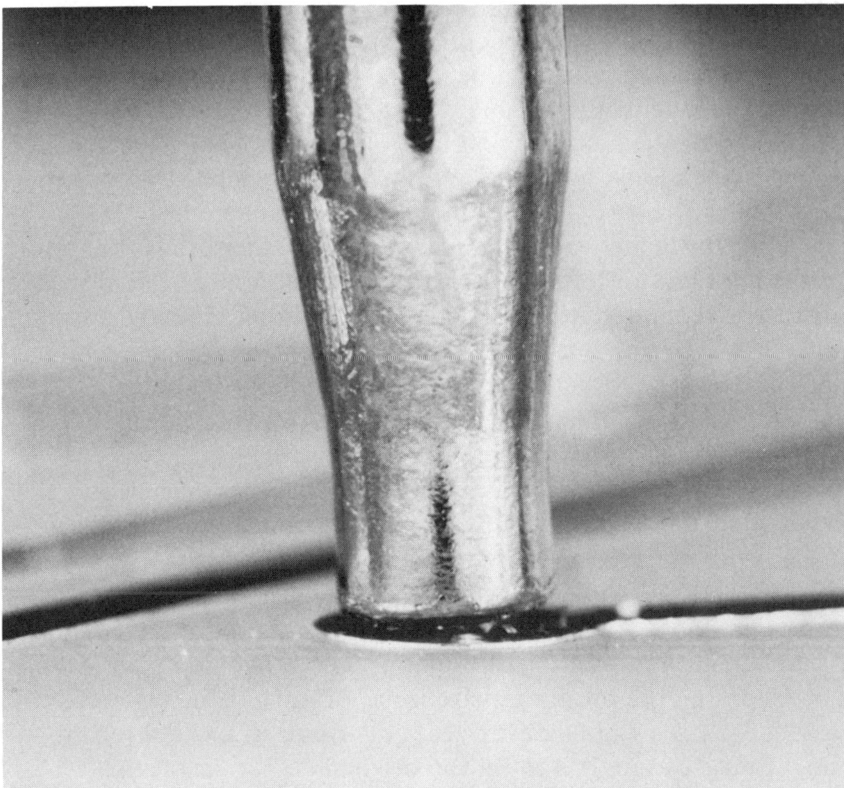

Figure 6.51 Heated vacuum tip overpad. (Courtesy of OMNI Training Corp., Cucamonga, CA.)

the product. The board must not be destroyed because all the work is for naught. Each component must be individually evaluated since testing may be required of the devices once they are removed and any damage to the leads would prevent this testing and troubleshooting from occurring.

Consider the weight and size of the tool from an operators perspective. The tool must feel comfortable in the hands of the operator and should be balanced for ease of use and maneuverability. From this author's perspective, the engineer should decide the power and vacuum requirements and define the tip size to be used. Then with inputs from the operators, let them decide which tool meets the engineering requirements and feels best to them. Remember, the operators will be using the tool every day so their inputs are needed and valuable.

Consider the heat it generates. The handle area when the tool is grasped must be cool at all times. If any heat or hot spots are evident, then the tool should not be recommended.

The workbench layout must address whether or not the operators are right- or left-handed and the tools should be so positioned to accommodate that preference. A holder must also be available to store the tool while it is not being used. This holder must be designed so that the tool is easily removed and installed and has the proper shielding to prevent anyone from burning themselves when touched.

The heat source and power must conform to local electrical codes. Consider vacuum sources for convenience and flexibility of the process. If air lines are convenient to existing workstations, then venturi-type vacuum systems can be considered. If air lines are not available, the choice is then limited to self-contained units like manual bulbs or ambient vacuum tools as shown in Fig. 6.52.

Choosing the proper tip size is important. If the wrong size is used, the pads will be destroyed, necessitating another repair. The purpose of repair is to fix the original defect, not cause another one. The tip should be sized as follows: the outside diameter (OD) (Fig. 6.53) should be no larger than the pad; the inside diameter (ID) (Fig. 6.54) large enough to accept the diameter of the lead into the hole.

How to use a solder removal tool The concept of the heat vacuum is to draw the solder out of the hole once it is molten. Additionally, the air flow that is created in the plated through hole after the solder removal phase cools the solder below the eutectic temperature. Therefore if the tip is still in contact with the lead, oscillating motion (Fig. 6.55) will allow the air to flow through the hole, solidifying the solder which will create a loose lead in the hole. Otherwise, the lead would solder itself to the wall of the plated through hole and the same process would have to be repeated to desolder the lead.

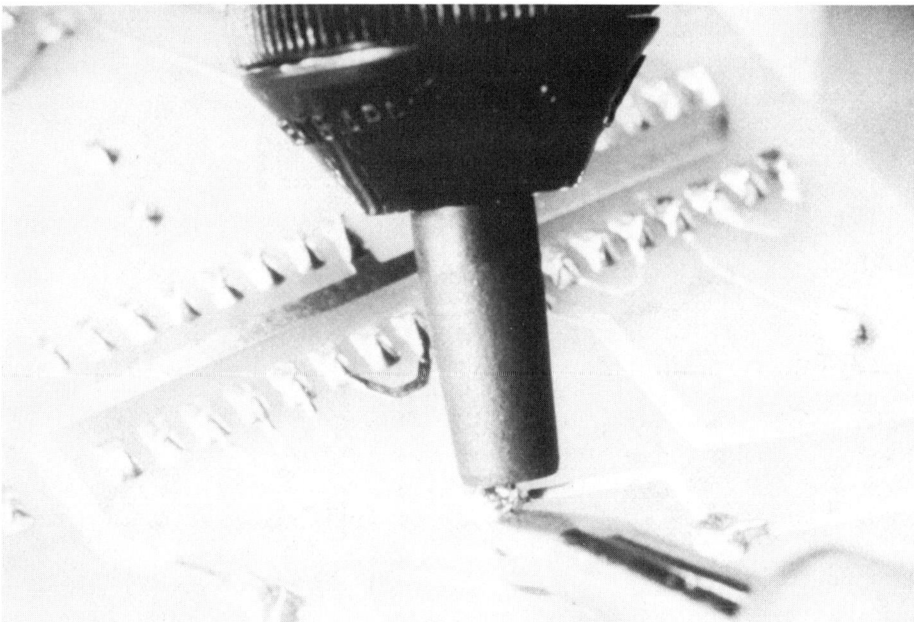

Figure 6.52 Solder iron and vacuum bulb. (Courtesy of OMNI Training Corp., Cucamonga, CA.)

When using the heated vacuum tool to remove the solder from the plated through hole of a double-sided board, no solder iron is necessary. However, when printed wiring assemblies start exceeding four layers and the inner layers are 2-oz. copper foil thickness, removing the solder from connecting plated through holes is difficult and could cause damage to the barrel and pad of the plated through hole. To prevent this damage, heat is simultaneously applied to both sides of the plated through hole with one of the tools being a heated vacuum tool and the other being a solder iron. Once the solder has become molten throughout the barrel, the vacuum tool is activated thereby removing the solder from the plated through hole.

Applying heat from both sides makes an assumption that accessibility exists to the component leads on both sides of the

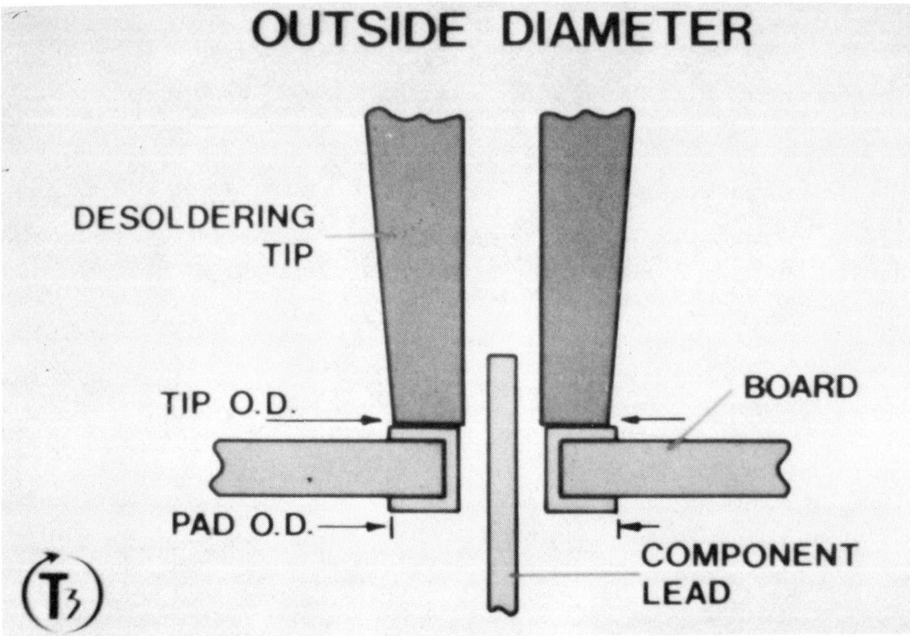

Figure 6.53 Outside diameter of desoldering tip. (Courtesy of OMNI Training Corp., Cucamonga, CA.)

assembly. The ambient-type tool is used solely with a solder iron. The iron is used to heat and melt the solder in the plated through hole, and then the vacuum tool is used to draw the solder from the hole.

Braid Removal

Braided copper wire that is fluxed and sometimes tinned is used for the braid removal process. Its use is very simple. Place the braid over the solder to be removed, apply the iron to the braid, heating it and melting the solder (see Fig. 6.56). Once the solder is molten, it will wick up into the braid, thereby removing it from the board. The braid comes in various widths and thickness and is simple to use.

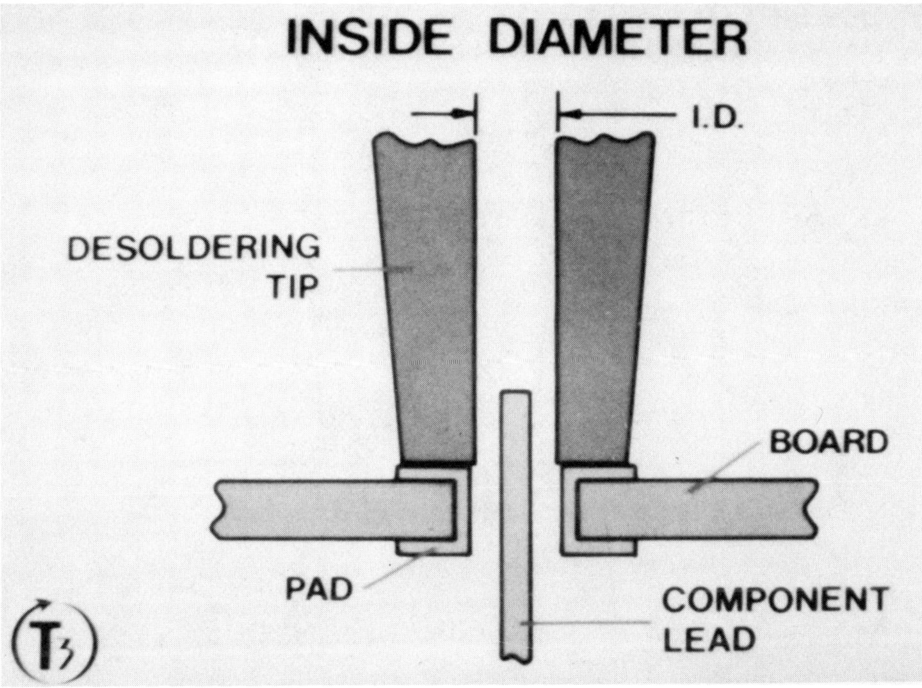

Figure 6.54 Inside diameter of desoldering tip. (Courtesy of OMNI Training Corp., Cucamonga, CA.)

Hot Air Reflow

Hot air reflow was designed for the removal of multileaded components (see Fig. 6.57). The concept of hot air reflow is to use localized hot air to reflow the solder and remove the device. To determine whether or not the proper temperatures have been attained, a thermal paste which melts at predetermined temperatures is placed on the device and in close proximity to the leads.

Using thermal paste which reflow at a similar temperature to the solder (e.g., 380°F) and assuming that the solder used is eutectic (63-37) or near eutectic (60-40), once the paste melts the solder should also be molten and the components are then ready for removal.

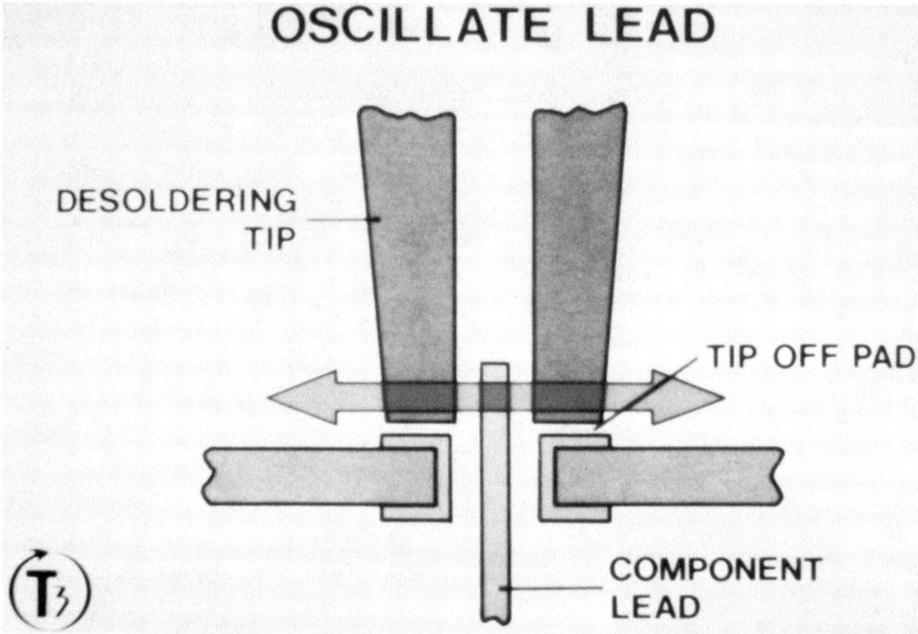

Figure 6.55 Oscillating the desoldering tip. (Courtesy of OMNI Training Corp., Cucamonga, CA.)

Proper training is necessary for using hot air reflow. The operator must understand its purpose and underlying concept. Improper training will lead to shortcuts that will destroy the assembly.

A disadvantage of the hot air repair method is its ineffective convection heat transfer process to melt the solder. This must be understood to prepare any manufacturing facility using this system. It takes approximately 1.0-1.5 min to reflow a large multileaded device (122-144 pins) while integrated circuits or dual in-line packages take approximately 30-45 sec. This is the way the system works and this time cannot be shortened by increasing either air flow or temperature. The devices and solder joints are being heated by a convection process and the ineffectivity of this heat transfer method is time consuming. Utilizing deflectors to

Figure 6.56 Solder braid. (Courtesy of OMNI Training Corp., Cucamonga, CA.)

direct the air flow to only the solder joints will improve cycle time and this must be investigated for each device.

The design of the orifice in the air tubes is not large enough to accommodate large devices and high pin count surface mount components. Therefore when these types of devices are being removed, circular motion beneath the air flow is necessary to accomplish unilateral heating across the entire surface of the device.

Surface mount components also require this type of equipment (Fig. 6.58) for component removal. However, the benefit of bottom side heating does not exist due to the absence of a through-hole connection. Therefore, all the heat must be applied from the component side. Additionally, the area beneath or directly opposite to the component may also be occupied by an-

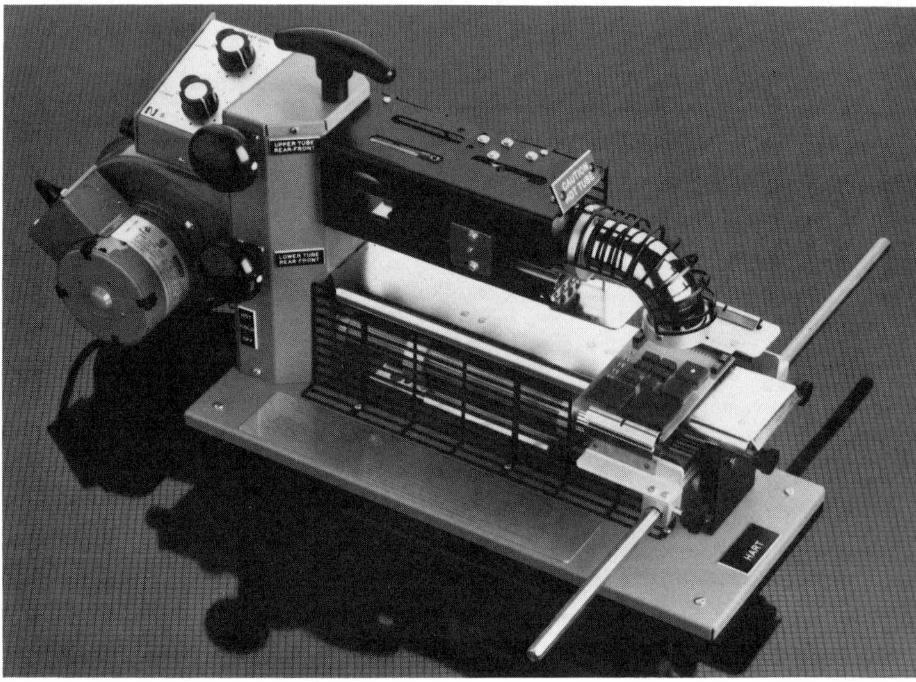

Figure 6.57 Hot air repair terminal. (Courtesy of Nu-Concept Systems Inc., Colmar, PA.)

other surface-mounted component. Hence this precludes that heat cannot be applied to side 2 of the assembly. If any heat is applied, it must be minimal, so as not to damage or reflow the components on the bottom side of the assembly. A major concern with surface mount components is the improper direction of the air flow when removing or replacing a device. Extraneous flow could reflow nearby components and potentially destroy the integrity of the solder joints.

Once the process has been designed and characterized for each component and assembly, the information must be documented in a process sheet detailing sequentially the steps necessary to accomplish the task. This must be done for every opera-

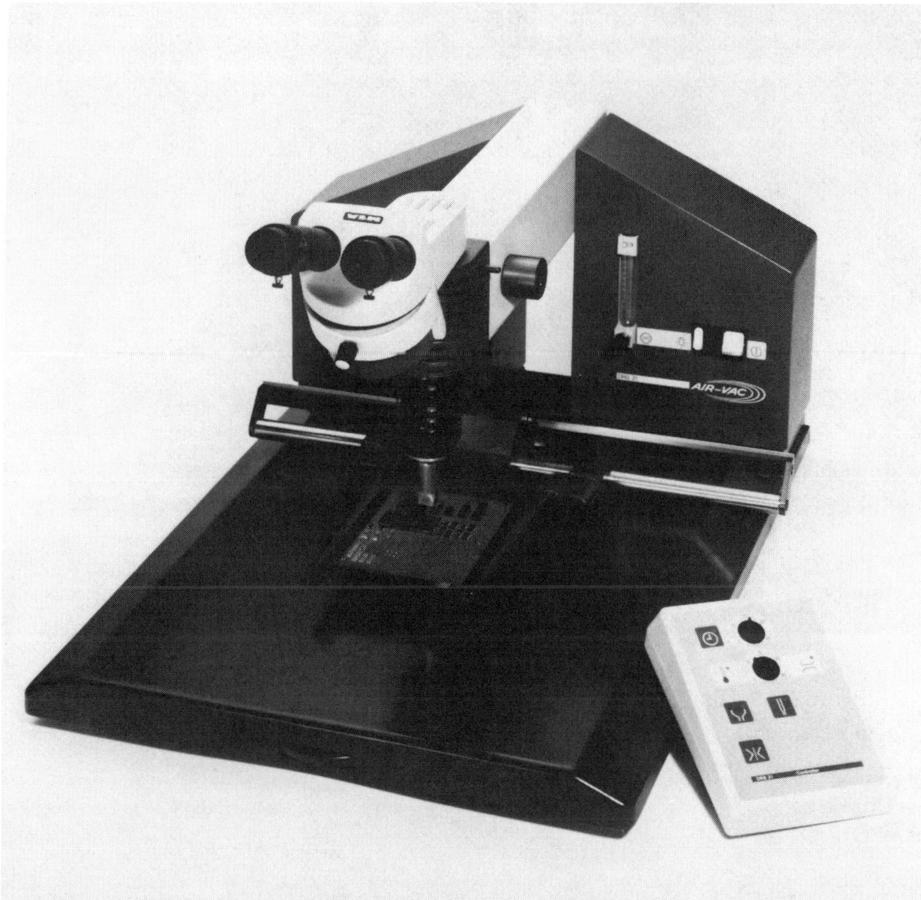

Figure 6.58 Air-Vac hot air removal tool. (Courtesy of AIR-VAC Engineering, Milford, CT.)

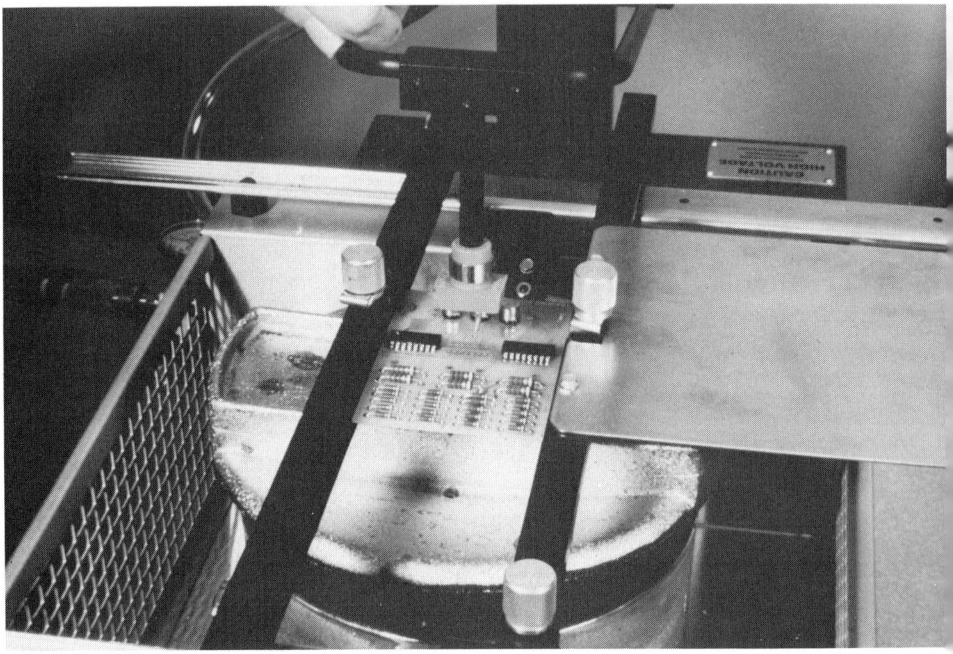

Figure 6.59 Board in fixture. (Courtesy of OMNI Training Corp., Cucamonga, CA.)

tion. Maintenance and repair procedures must be documented to aid in reducing and/or eliminating downtime of the equipment.

Hot Flow Solder Wells

Operating this type of system requires some basic assumptions.

1. Solder pot alloy must be fixed, preferably 63-37 to QQ-S-571E.
2. Flux for fluxing component leads is defined.

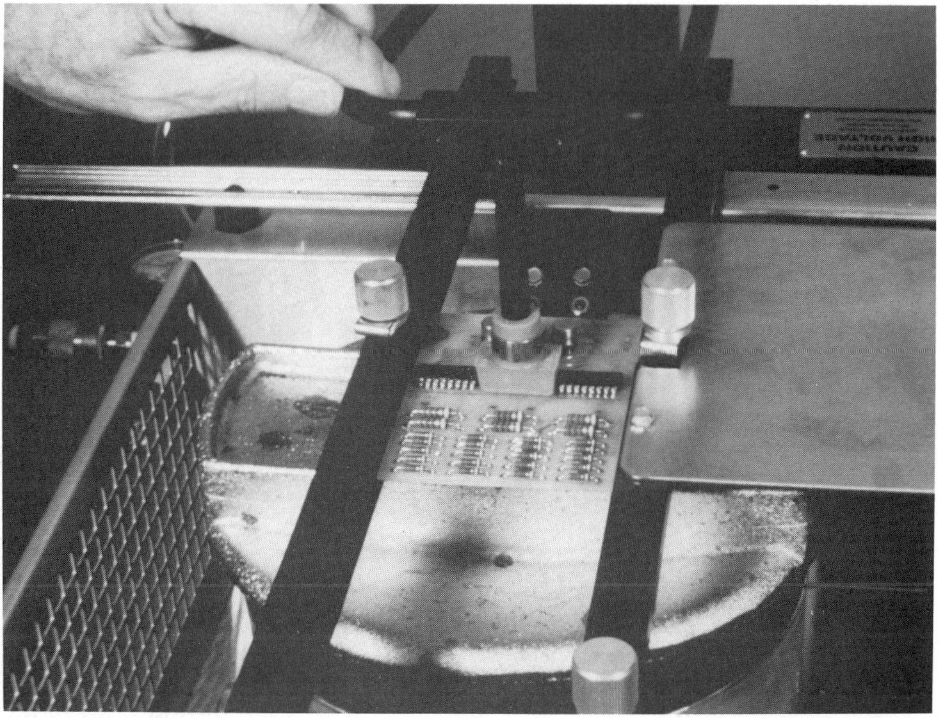

Figure 6.60 Rubber well located over component for locational purposes. (Courtesy of OMNI Training Corp., Cucamonga, CA.)

3. Solder pot temperature fixed at 495 ± 5°F.
4. Lead outline is defined and solder wells are available for each lead outline.

Procedurally, flux area of component, position board onto fixture (Fig. 6.59), define location of component to be removed (Fig. 6.60), grasp component with special tool (Fig. 6.61), and activate machine. When solder reflows, remove component (Fig. 6.62).Also do the following: (1) Certify that the leads are not clinched flat to the bottom of the assembly. (2) Certify that the solder is completely reflowed prior to removing components.

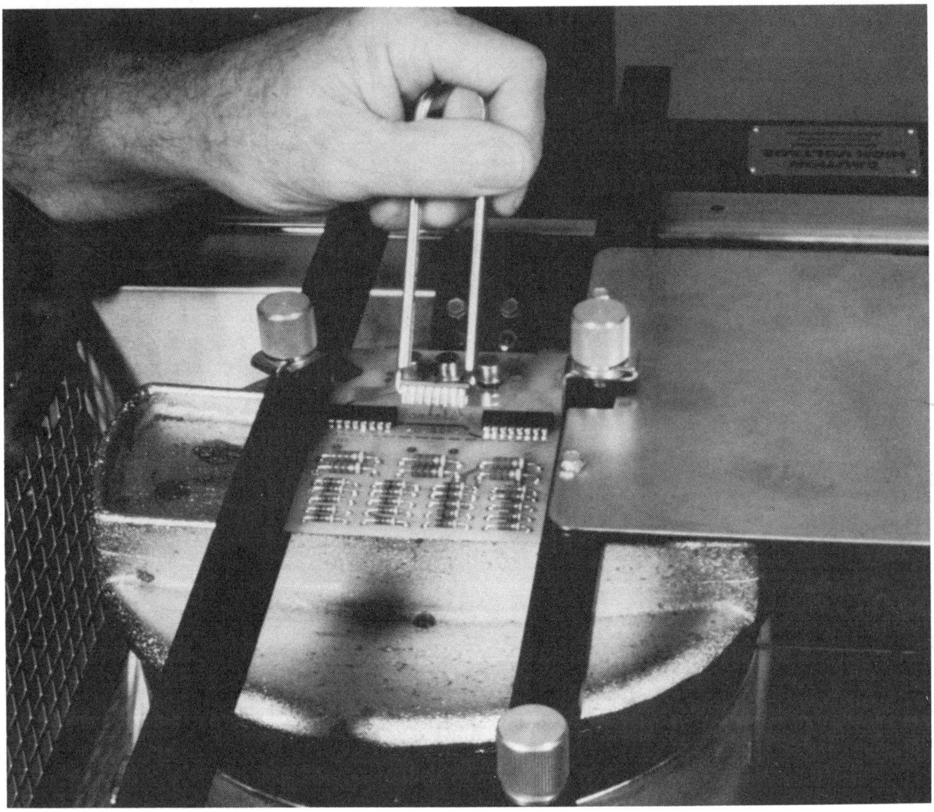

Figure 6.61 Grasping component with removal tool. (Courtesy of OMNI Training Corp., Cucamonga, CA.)

(3) Know the board layup to certify how much heat and how many cycles are needed to remove the device. (4) Identify flow rates and cycle time for each component.

Hot flow-well systems are the most efficient of the removal tools for multileaded devices. The heating method is conduction. Molten solder is brought into contact with the assembly, the area being defined by the flow well which outlines the components. As the solder contacts the leads and plated through holes, the heat is immediately transferred to the components reflowing the solder in the plated through holes.

Figure 6.62 Component removal. (Courtesy of OMNI Training Corp., Cucamonga, CA.)

As the solder becomes molten on the component side, a special tool which grasps the component should be used to pull the device off the assembly. Once the component is removed the pumping of the well is stopped removing the contact with the assembly. At this time, there is still enough residual heat in the board to keep the solder molten. In some systems there is a timed quick air blowoff unit which directs air by a rubber cover of the same configuration as the solder well and this quick burst of air removes the molten solder from all the plated through holes (Fig. 6.63). This usually works on all holes except for plated through holes that are connected to the inner layer ground and

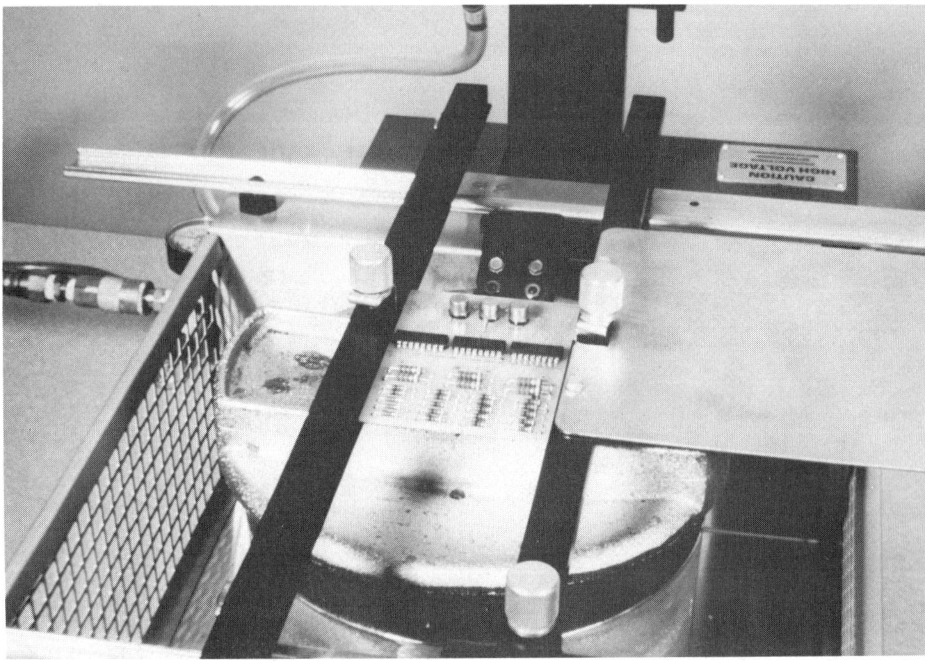

Figure 6.63 Removing molten solder from plated through holes. (Courtesy of OMNI Training Corp., Cucamonga, CA.)

voltage layers. Hence, to clean these holes, heated vacuum tools are used.

Maintenance of the hot well system The mechanism of pumping the solder is simple. Old equipment pumped the solder through a piston mechanism with a finite quantity of solder. There was enough caloric energy in the slug of solder being pumped to reflow the joints of the components in question. If there was not enough heat, the process was simply repeated. This was acceptable due to the small, multileaded components being removed.

However, with the large, multileaded components, such as pin grid arrays, large multileaded sockets, etc., the old systems did not have enough caloric energy to reflow all the solder joints.

Figure 6.64 Small solder well. (Courtesy of OMNI Training Corp., Cucamonga, CA.)

Therefore equipment changes were required where more solder could be pumped on a continuous basis (Figs. 6.64–6.66). This kind of system has a larger solder pot than the old systems and can be used for multipurpose components removal equipment.

These systems are prone to the same maintenance as solder wave equipment. Pumps need lubrication and dross buildup must be controlled. Also the metal elemental contamination levels must be checked to certify that good solder joints are made with the proper metallic alloy.

Figure 6.65 Medium solder well. (Courtesy of OMNI Training Corp., Cucamonga, CA.)

Figure 6.66 Large solder well. (Courtesy of OMNI Training Corp., Cucamonga, CA.)

6.5.8 Component Replacement

Component replacement systems are simplified because of all the holes being free of solder. When replacing components, place the component in the board (Fig. 6.67), flux the area, activate the machine, and solder all the leads (Fig. 6.68). The final steps of replacing components are as follows: (1) check the quality of the solder joints. (2) Remove the flux residues. (3) Certify the electrical parameters of the product.

Figure 6.67 Reinserting component. (Courtesy of OMNI Training Corp., Cucamonga, CA.)

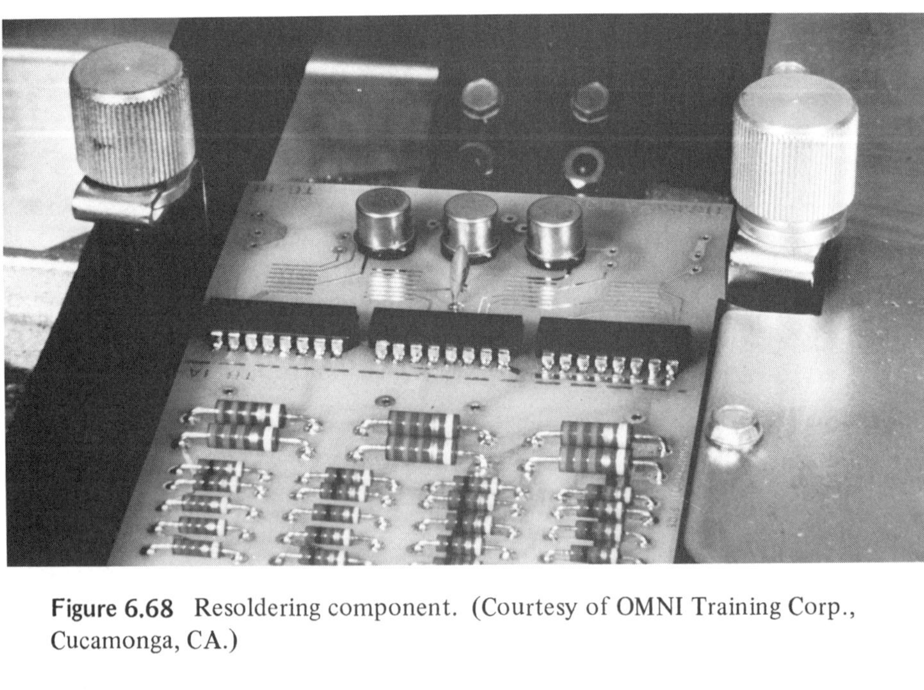

Figure 6.68 Resoldering component. (Courtesy of OMNI Training Corp., Cucamonga, CA.)

ASSOCIATED READINGS

1. Richard E. Spencer, *Hand to Hand Combat*, Circuits Manufacturing, May 1986, pp 65-69.

2. Walter H. Schwartz, *Soldering Irons: Still Hot*, Assembly Engineering, March 1986, pp 34-37.

3. *Solder/Cleaning, Test and Rework in Electronic Assembly*, Assembly Engineering, May 1985, pp 66-73.

4. Prayag Prasad, *What To Do About Hand Soldering Damage to Printed Circuits*, Insulation/Circuits, November 1982, pp 89-92.

5. Mervin Baranick, *Cored Solder For Aluminum and Aluminum Alloys*, Assembly Engineering, November 1982, pp 22-23.

6. Joe Keller, *Lited Plated Through Holes–Good or Bad?* Assembly Engineering, August 1982, pp 16-17.

7. Prayag Prasad, *Contributing Factors for Thermal Damage in PWB Assemblies During Hand Soldering*, IPC-TP-394, IPC Fall Meeting, Dallas, September 1981.

8. Harold F. Vandermark, *Desolder and Remove DIP's in One Motion*, Insulation/Circuits, August 1976.

9. Richard O. Johnson, *Modern Hand Soldering with Electric Soldering Irons*, Electronic Packaging and Production, July 1976, p-26.

10. Howard Manko, *Printed Circuit Touch-Up*, Circuits Manufacturing, July 1978, pp 83-88.

11. *Upgrading Soldering Quality: Presoldering Operations*, Production Soldering Clinic, Insulation/Circuits, August 1979, pp 17-19.

12. *How to Make Soldered Connections to Terminals*, Production Soldering Clinic, Insulation/Circuits, March 1980, pp 60-74.

13. Ref. 12, June 1980, pp 33-38.

14. Mark Rowland, *Evaluating Hand Soldering Irons*, Assembly Engineering, December 1984, pp 26-27.

15. Richard N. Buggle, *Guidelines for Qualifying and Certifying Soldering Irons*, Insulation Circuits, April 1978, pp 23-26.

16. James Bausell, *Printed Circuit Technology Reports*, Electri-Onics, February 1984, p. 37.
17. J. Ferry, *Developing and Implementing a Successful PCB Repair Program*, Electri-Onics, January 1984, pp 21-23.
18. M. F. Comerford, *Removing Solder Bridges from Printed Circuit Boards*, Insulation/Circuits, April 1981, pp 31-33.
19. Howard W. Markstein, *Ensuring Successful P. C. Board Desoldering*, Electronic Packaging and Production, May 1981, pp 53-67.
20. Michael E. McBride, *The Wicking Method of Solder Removal*, Insulation/ Circuits, December 1980, pp 25-28.
21. Clyde M. Vipond, *Part I: Wire Stripping Facts and Tools*, Insulation/ Circuits, August 1978, pp 17-21.
22. James P. Langan, *Tips of Selection of Wire Solder For Manual and Automatic Soldering*, Insulation/Circuits, December 1977.
23. *Soldering Iron Techniques and Maintenance*, Production Soldering Clinic, Insulation/Circuits, June 1979, pp 20-30.
24. J. A. Sylvester, *Hand or Manual Soldering?* Assembly Engineering, July 1976.
25. Howard Manko, *Tune Up Your Hand Soldering*, Assembly Engineering, July 1976.
26. Joseph A. Sylvester, *Matching the Iron to the Job*, Electronic Packaging and Production, March 1968.
27. Richard O. Johnson, *Soldering Tools–Thermal Output Regulation*, Electronic Packaging and Production, July 1978.
28. Joe Keller, J. Dennis Keller, *Temperature Controlled Soldering Irons*, Circuits Manufacturing, September 1976.
29. Casimir D. Bernard, Joseph A. Sylvester, *Soldering and Components: How and Why to Measure Effects*, Insulation Circuits, January 1974.
30. Joseph A. Sylvester, *Soldering Problems/Solutions: A Collection of Recent Case Historians*, Insulation/Circuits, November 1969.
31. Larry Lapanne, *Hand Soldering: A Look Ahead*, Quality Magazine, January 1978.

32. Nicholas Rusignuolo, *Nondestructive Hand Soldering of Microelectronics*, SMS Technical Paper EE77-117.

33. Gert Becker, *Specifying and Measuring the Working Capacities of Soldering Irons and Tips*, Assembly Engineering, October–November 1975.

7

Surface Mount Process

7.0 INTRODUCTION: WHAT IS SURFACE MOUNT?

Surface mount, as the name suggests, refers to the method by which components are assembled onto printed wiring assemblies. These components are mounted onto the surfaces of the printed wiring board at predetermined locations on prepared solder pads or land areas. For consistency, pads will now be called land areas. The components can be mounted on either side 1 or 2 or both of the printed wiring assembly.

The main differences between surface mount and plated through-hole assemblies are the methods and processes used to assemble components into printed wiring boards. Surface mount components are positioned onto prepared land areas, while plated through-hole components are placed intrusively into plated through holes in the board itself. Plated through-holes components are mechanically fastened whereas the surface mount devices are merely placed into or, in some cases, onto the solder paste; therefore handling of the product is much more critical to prevent component movement and misalignment.

Surface mount processes create the necessity for more controls, especially in the handling aspects of the product throughout the process. The major confusion in surface mount technology relates to the types of components used, and these are usually categorized into two styles: active and passive. Passives are identified as resistors and capacitors while actives are integrated circuits, transistors, memory devices, gate arrays, PROMS and ROMS, etc.

The conversion from through-hole technology to surface mount technology is increasing rapidly for a myriad of reasons, the largest being the savings of space on printed wiring boards. Although this was the original intent, increased functionality of components and products are also pushing the technology. Hence the two major thrusts for conversion are space saving and increased functionality. Both have special assembly process requirements.

7.1 PRODUCT DESIGNS

The product must be designed for surface mount assembly. The implication is that many elements must be reviewed and defined to certify and verify producibility and manufacturability. The following must be defined:

- Land areas and land configuration, length, and width
- Component lead configuration
- Component lead plating
- Component styles and sizes
- Board material, laminate, plating composition and thickness
- Whether components will be used on one or both sides of the assembly

- ○ Whether the assembly will be mixed mount technology
- ○ Purchase and in-process specifications for product compliance

This preparatory work must also include reliability testing for all manufacturing variables. This is necessary to understand the reasoning behind the assembly and process specifications.

7.2 SURFACE MOUNT PROCESS FLOW

Each of the manufacturing steps of the surface mount assembly process must be complementary to each other for producibility and manufacturability certification. Three surface mount assembly sequences are described below.

7.2.1 Type 1 Assembly

Type 1 assembly relates to surface mount assembly only, which can be on either surfaces, side 1, side 2, or both. See Fig. 7.1. This process can be used for any style of component, i.e., active or passive. If side 2 components are necessary by design, repeating the above process will satisfy the requirements.

Depending on the product requirements, adhesives may be necessary for attachement of side 2 components. Assemblies with components on both sides must be designed to certify that the component weight will not overcome the tensile strength of molten solder because if this condition occurs, the components will fall off. If in fact the components are too heavy and it is suspected that they will fall off, they should be mechanically secured with an adhesive.

In choosing the adhesive consider the coefficient of expansion of both the component and the board material and also its resiliency in the repair cycle. It must have the ability to be soft-

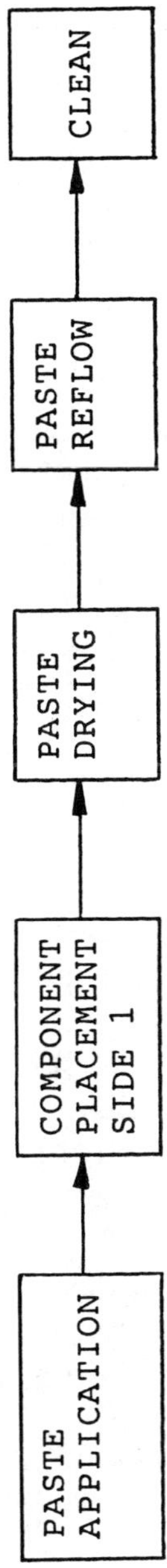

Figure 7.1 Type 1 assembly flow chart.

ened by heating and therefore breakable, allowing the removal of the component.

7.2.2 Type 2 Assembly

Type 2 assembly relates to plated through-hole components and passive surface mount components. The through-hole components are side 1 and the passive surface mount components are side 2. See Fig. 7.2.

For passive devices on side 2, this is a high-volume process. No paste deposition is required with its associated quality requirements. The process has a drawback, however, which is the entrapment of flux residues beneath the passive devices. These components are leadless and usually within 3 mils off the surface of the laminate. This is enough space for entrapment of wave flux. If the wave flux is corrosive, byproducts will create electrical malfunctions and long-term reliability problems with degradation of the physical solder joint.

7.2.3 Type 3 Assembly

Type 3 assembly is defined as plated through-hole components on side 1 and surface mount components on both sides 1 and 2. See Fig. 7.3 for side 1 assembly. If side 2 passive components are used, then the flow would be as shown in Fig. 7.4. Side 2 passives are mounted after the plated through-hole process to certify that the cut and clinch mechanism does not damage the surface mount components. This must be considered in the design layout as to where the passive components are to be mounted in relation to the footprint area of the side 1 components. Therefore identify the footprint area occupied by the cut and clinch mechanism of the automatic insertion equipment to prevent damage to surface mount components during automatic component assembly.

The sequence of assembly and the relationship between each step cannot be overemphasized.

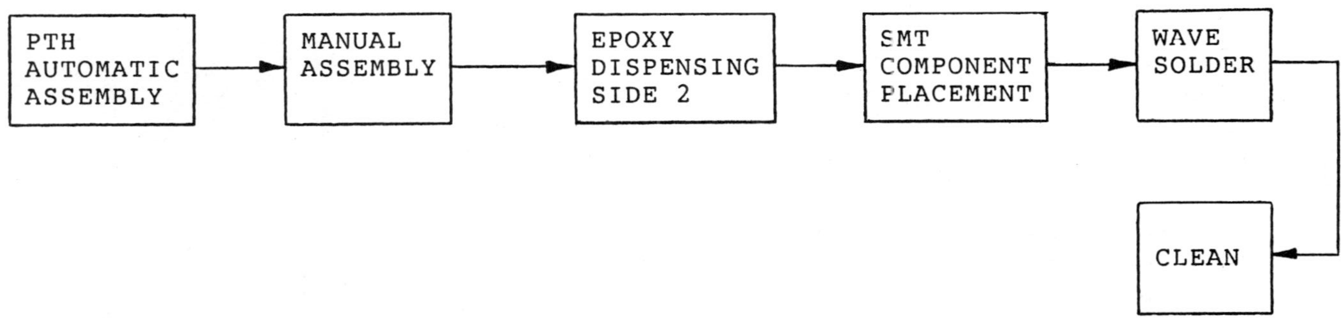

Figure 7.2 Type 2 assembly flow chart.

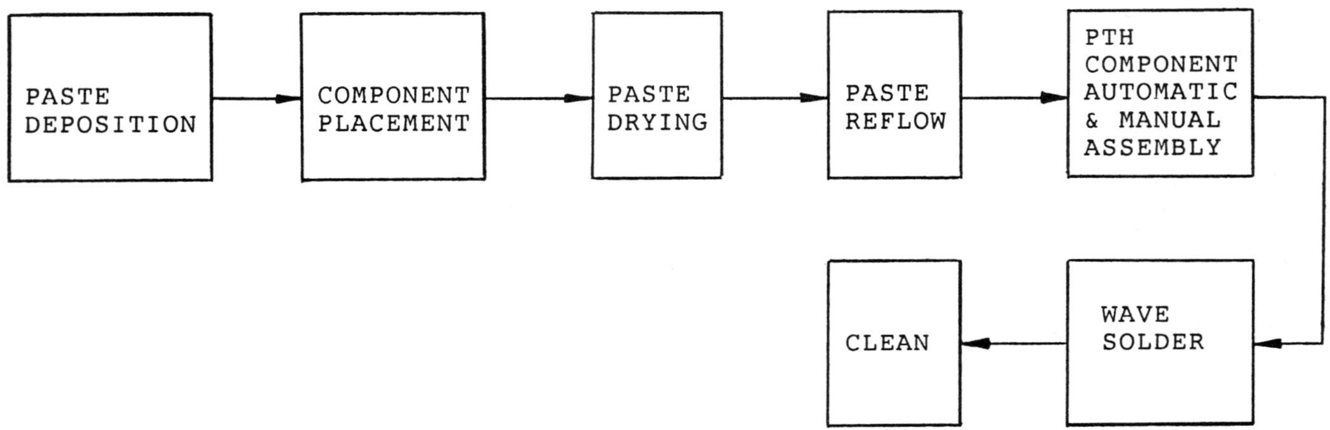

Figure 7.3 Type 3 assembly flow chart #1.

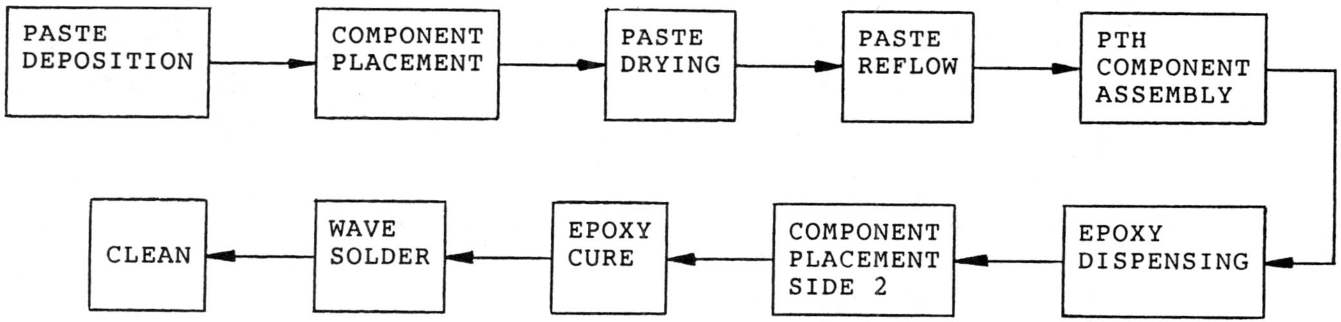

Figure 7.4 Type 3 assembly flow chart #2.

7.3 MATERIALS USED FOR SURFACE MOUNT ASSEMBLY

7.3.1 Solder Paste

Solder paste or solder creams are used for increasing the amount of metal on the board to create the solder joint between the components and printed wiring assembly. In comparison, intrusive-type components use solder from the wave solder systems to create those types of solder joints. Additionally, various solder paste alloys are available for particular applications. Some applications are (1) low melting temperatures, (2) prevention of silver leaching and/or dissolution, (3) high strength, (4) high ductility, see Table 7.1.

To properly understand solder paste or creams, the following questions must be answered:

Table 7.1 Paste Alloys

Alloy Composition				
Sn	Pb	Ag	In	Application
63	37	–	–	For printed and flexible circuitry
60	40	–	–	For printed and flexible circuitry
62	36	2	–	Design for thick-film, silver-based metalization, prevents leaching of silver
96.5	–	3.5	–	Thick-film circuitry, superior wetting, and good strength
10	90	–	–	Bottom side assembly to prevent reflow during wave soldering, chip bumping
–	50	–	50	Prevents solution of gold plating

Source: Adapted from the following technical bulletins: Fusion Incorporated Technical Bulletin "Solder Paste" A-112; Fusion Incorporated Bulletin "Screenable Solder Paste" A-110; Indium Corporation of America Technical Bulletin SMT-1-3M, 9-84; Alpha Metals Inc. Technical Bulletin "Solder Paste."

- What are they?
- How are they specified or chosen?
- What are the important parameters to measure and record for process control?

Solder paste is composed of spherical and elongated solder powder of given shapes and sizes. These solder particles are then segregated according to their size, which is accomplished by sifting the powder through a specific mesh size screen with controlled opening dimensions through which the correct size particles will pass through the larger ones will not. See Table 7.2. The powder is then combined with a liquid system which contains flux and a viscosity adjustment material, and it is suspended within this liquid system. The flux material can be of any type, such as rosin or organic acid base.

Paste Classification

Solder pastes are usually specified by defining:

- Quantity of solder
- Mesh size which determines the particle size
- Viscosity
- Type of flux system
- Flux activity levels

A typical standard classification is:

- 88-2-96 RMA
    - 88% by volume of metal content
    - 200 mesh screen
    - 96 KCPS viscosity
    - rosin mildly activated flux

Table 7.2 Mesh Count vs. Wire Diameter vs. Typical Mesh Opening

Mesh count	Wire diameter	Mesh opening
80	0.0020	0.0105
80	0.0037	0.0088
80	0.0055	0.0070
105	0.0030	0.0065
120	0.0026	0.0057
200	0.0016	0.0034
200	0.0021	0.0029
325	0.0009	0.0022
325	0.0011	0.0020

The quantity of the liquid system in relationship to the total prepared volume of solder paste is defined by weight percentage, i.e., 88% solder paste contains 88% metal and 12% nonmetal liquid system.

Paste Volume

When designing the solder joint, the rectangular area of the pad and the required height of the solder joint determines the volume of metal to be deposited as paste. The rectangular area of the land is identified as the length and width of the pad. When the correct height of the finished solder joint is defined, the proper volume of paste can be applied to the land area by controlling the height of the deposit.

The volume of the reflowed metal is different from the weight measurement characteristics made earlier. In its original form, the solder spheres occupy a much larger area due to the individual solder spheres and the area each sphere occupies. When the solder paste is reflowed, the solder in the paste forms a homogeneous mass which is denser, and although it weighs the same it occupies much less space and hence has a smaller volume.

Incoming Inspection of Solder Paste

This operation cannot be stressed enough to certify that solder paste conforms to the requirements established by the particulars of the process. The quality of solder paste cannot be compromised and accepted if all the inspection line items are not met. If it is compromised and accepted, defects will result such as open solder joints and solder shorts. In the assembly process these types of defects are more costly to repair than to prevent, especially if good solder paste had been used initially.

Incoming inspection tests must be performed on all parameters of the solder paste:

Element	Test method
Sphere size	Fineness of grind test
Flux activity	Resistivity test
Flux quantity/metal quantity	Weighing test
Slump	Visual measurement
Viscosity	Viscometer readings
Solderability	Reflow and measurement
Spread	Reflow and measurement

Screening or stenciling the solder paste onto the pwb will determine which sphere size will be used in the solder paste. Improper sphere size will prevent good penetration of the solder paste through the screen or stencil. This results in improper quantity of solder paste on the pads.

Flux activity if not correct will potentially create some electrical malfunctions if it is not completely removed during the cleaning cycle. It will impact the insulation resistance of the circuitry.

Flux and metal percentages in the solder paste can impact the reliability of the solder joint. Product design requirements must define reliability criteria, and deduced from this information is the quantity of metal required to make the final solder joint interconnection.

Slump is a term used to identify the ability of the solder paste to remain where it has been placed. A liquid system when placed on a surface will tend to flow outward because it has no definite shape or size. The deposited paste must not be allowed to slump, i.e., it must remain as deposited without losing any of its physical characteristics of height, width, and length. If any deviation occurs from the original form deposited, it is said to have slumped. Slump is allowed to a certain extent and the requirements usually define how much deviation can occur from the original width of the deposit.

Viscosity is a term used to measure the flow characteristics of a fluid. It can be stated as the relationship between the shear rate and shear stress of a material or a measurement of the forces acting on an object displacing the material. In solder paste, it is a very ambiguous measurement to make and understand. Many studies are being performed to determine the relationship of viscosity to the slump characteristics of the solder paste. It appears at present that solder paste which fails viscosity tests will also fail slump requirements. The important criteria, however, is the slump. The paste must stay where it is placed and the physical dimensions of the deposit must be changed. This is a typical characteristic of thixotropic materials.

The term *thixotropic* is defined as the property of a liquid to resist flow unless acted upon by an outside force. This means that when measuring the viscosity of the solder paste it is important to determine whether or not the solder paste has been stirred before the test is performed. If it is stable and has not been stirred or mixed within a couple of hours it will have a high viscosity reading. When the solder paste has been stirred the viscosity is reduced and the measurements reflect such changes.

For good solderability the paste must be solderable. When reflowed it must not create solder balls or spatter but create one homogeneous form of solder. Spread implies the ability of the solder paste to wet the surface of the elements being soldered. It is complementary to the activity of the flux.

7.4 SOLDER PASTE DEPOSITION

What are we looking for from deposited paste? Once the paste is in place, it must stay there until the components are placed into it and it is reflowed. Tackiness is required of the solder paste to act as an adhesive for component anchoring prior to reflow.

The ability of the paste to be screened or stenciled is determined by the viscosity of the paste. This dimension is in KCPS or thousand centipoise.

Another element to be considered with solder paste is its ability to reflow and not create any solder balls. To create a homogeneous solder slug, all the solder spheres must reflow simultaneously. This is important because of the amount of oxides which are available on the surfaces of all the solder spheres, not including the component lead and land areas. Large spheres occupy more space in the paste; therefore, usually have less oxide. Small spheres are greater in quantity and therefore more oxides are evident.

Hence, all of these elements of solder paste are collectively called the rheology of the paste. Rheology is defined as the science of deformation of bodies under stress and it encompasses characteristics such as flow, elasticity, yield points, and viscosity.

7.4.1 Deposition Methods

Once the paste has been chosen, the method of application must be defined: screen or stencil. If screen is used, the sphere size becomes a critical issue. Spheres which are large and openings which are small will prevent the proper dispensing of the paste. Also the quantity of threads in the mesh which determine the size of the opening are not consistent. Therefore, consistent deposition is sometimes suspect with the wrong paste size and screen mesh size.

One method to overcome this is to use stencils, either brass or stainless steel. The consistency of deposit is available with the proper choice of paste and opening in the stencil. The stencil fabricator must certify the technique of making the openings in the stencil material. The edges of the openings must be straight and smooth to allow the solder paste to flow through the material itself. If the edges of the openings are rough or not perpendicular to the product being stenciled, improper solder paste deposition will occur.

7.4.2 Selection of Deposition Methods

Paste deposition screen printing vs. stenciling vs. mass spot application: which is best? Depending on the volume requirements, one or more of these can be used for dispensing paste. Initially, mass dispensing could be performed by screening or stenciling. Which one to use depends on the paste rheology and quantity of paste needed to make the particular solder joint (see Figs. 7.5-7.7). Equipment vendors have all the information necessary to properly determine which system best suits any particular application. Consideration must be given to accuracy of deposit and repeatability of the system. Other major concerns are board size being screened or stenciled, whether or not multi-up methods are used, and ease of adjustment. Ease of adjustment is of major importance when many different board types are being worked on.

Upcoming computerization of the systems will ease the task of alignment. However, its cost may be exhorbitant and out of reach for many people. Therefore, the system chosen should be one that is simple to operate and set up.

Attention to the size of the screening area and the consistency of the solder paste deposits across the entire board area are of major importance for good homogeneous solder joints. The thickness of deposit is based on the thickness of the screen or stencil and the snap-off distance, which is identified as the set distance from the bottom of the screen to the surface of the board.

The uniformity of the screen emulsion thickness is at time suspect across the entire screen area, especially if a thick 0.005-in. emulsion or greater is needed on the screen mesh. This is difficult to manufacture and should be worked with screen manufacturers to certify they in fact can fabricate the emulsion to such exact requirements.

Figure 7.5 AMI Presco Model 873. (Courtesy of AMI, North Branch, NJ.)

Figure 7.6 AMI Presco Model 2405. (Courtesy of AMI, North Branch, NJ.)

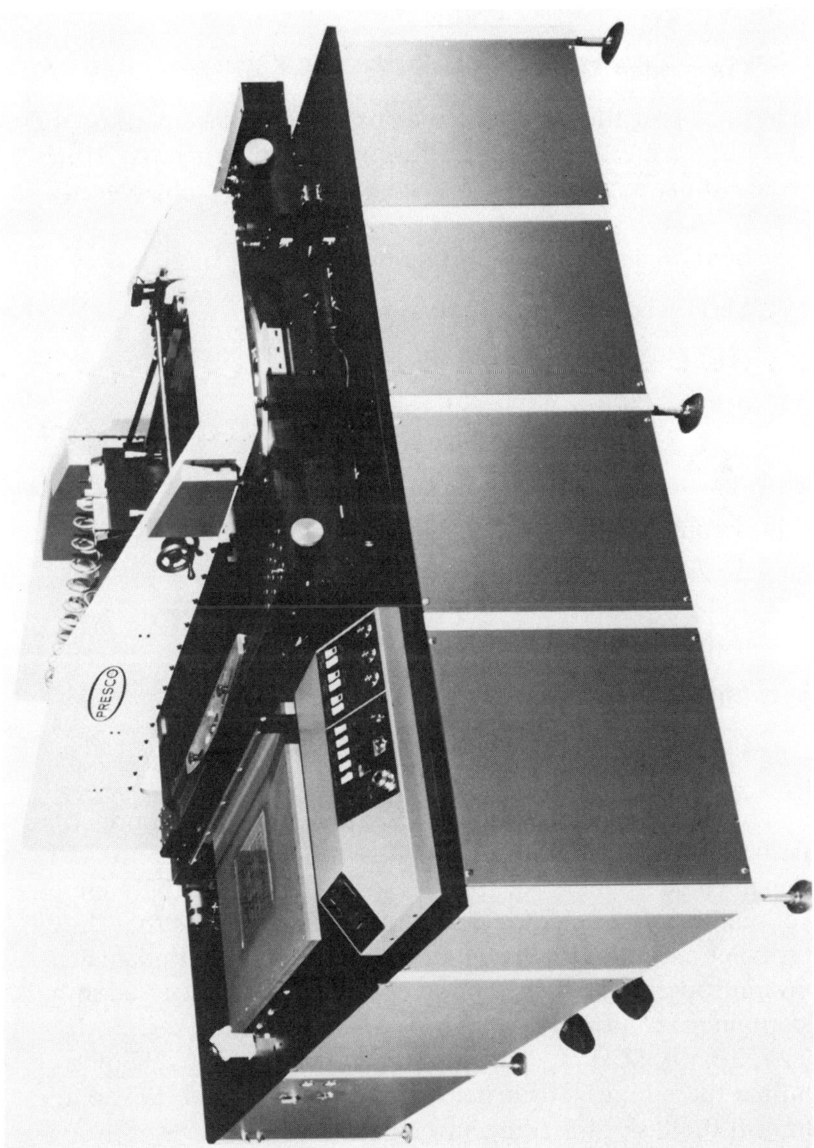

Figure 7.7 AMI Presco Model 2405. (Courtesy of AMI, North Branch, NJ.)

7.5 COMPONENT PLACEMENT

7.5.1 Component Types

The next step of the process is the component placement operation. Various methods are used to position components. However, one must consider the style of components being placed:

- SOIC small-outline integrated circuit
- SOT small-outline transistor
- MELF metal-electrode face-bonding
- PLCC plastic-leaded chip carrier
- CLCC ceramic leadless chip carrier

The lead configuration outlines are:

- J bend
- Gull wing
- Butt joints
- Lap joints

The methods of handling surface mount type components must be thoroughly analyzed. Surface mount components were designed to be smaller, occupy less space, and potentially increase functionality of the electrical assembly. The point is that they are smaller and therefore they are more susceptible to damage from improper handling methods. They can be purchased in bulk or containerized in various tapes of different width, e.g., 8, 12, 16, 24, 32, 44, 56 mm. The equipment must be capable of handling these tapes plus other handling systems in order to accomplish the task of picking and placing these devices. Surface mount components also mount closer to each other as compared to the standard through-hole components. Because of this, equipment must be designed to grasp the component from its top surface, centering it with some mechanized chucks, then positioning the device onto its required land configuration.

7.5.2 Automatic Placement

Automatic placement creates another problematic condition: placing surface mount components exactly onto the etched land areas. In through-hole component processes, the printed wiring board was oriented on fixtures by the tooling pins, from which X- and Y-coordinate corrections were made to some reference holes which were drilled during the same operation as the component holes. Hence a method was available to rezero each board in relation to the tooling holes. This methodology provided the accuracy necessary to reduce the quantity of misinserted parts and prevented damage to those parts. Another tool used for ease of through-hole insertion was the hole size utilized in the printed wiring board. The larger the hole the easier it was to insert the component.

With surface mount components, however, the accuracy of the placement must be determined, not by any drilled holes but by the location of the land areas. Since the fabrication process for aligning the artwork to the printed wiring boards is not an accurate process, the location of each land area within a pattern is more accurate than the location of the pattern to the overall area on a printed wiring board. Therefore, from board to board, the patterns could be off by as much as 10-12 mils. This is quite high relative to the accuracy needed to place the components on the land areas. If the pick-and-place equipment is as inaccurate in its component placement cycle as the overall etch pattern is to the printed wiring board, components will end up being placed onto the wrong land areas altogether. Therefore, a sensing mechanism must be designed to define the exact land location in relationship to some base zero location. This can be accomplished by a vision system and fiducial characters on the board or assembly.

Fiducial characters or marks must be part of the etch pattern and when viewed with the vision system, correlation or adjustments can be made for the correct land locations. As the component lead pitch distances becomes finer and finer (50-mil pitch on typical components and future components will have 25-12 mil pitch distances), any equipment designed for future application must take these elements into consideration.

This tighter accuracy requirement also becomes critical for pick-and-place equipment. Mechanical machine tolerances must be capable of self-correcting to certify accurate placement of any and all components.

7.5.3 Manual Placement

One area of concern is the low-volume manufacturers where the high cost of sophisticated equipment will not be available. How will components be accurately placed if there is only one or two per board? The only practical way to approach this problem is with fixturing. Fixtures can be accurately designed and fabricated that will match the accuracy of placing components on the required location. These fixtures can be fabricated from any of the standard type of materials available, provided they can be accurately machined.

They can be used as transfer plates, where components will be assembled into the fixture, then the prepasted board will be positioned over the transfer plate and the two will be brought together. Once they have contacted the transfer plates will be removed, leaving the component on the land areas of the printed wiring assembly. This method will suffice for devices with 0.050-in. pitch lead spacings. For 0.025-in. pitch lead spacing and finer, more sophistication will be required due to both the close proximity of each lead but also the fragility of the leads. If poorly designed handling methods are incorporated into the process they will be evident at this point, with poor lead registration.

Hence, in the case of low-volume/low-mix or high-mix/large-lead count PLCC components with 0.025-in. pitch lead spacing, the use of either special fixturing or modifying some piece of equipment with a grasping mechanism and a microscope will be necessary. This grasping mechanism can then be self taught for each particular board type and component location. This system can also be used for manual assembly in repair operations.

7.6 BAKE AND CURE

This process may not always be necessary, depending on the paste and the reflow process used in the product fabrication. Certain pastes have rheology characteristics which encompass special materials that do not evaporate during the product use or volatize quickly when subjected to reflow conditions. The process msut be designed based on certain constants, and when these constants are material-related the equipment must then be qualified per the material. This is in comparison to a process designed around the equipment where the process materials mus be specifically designed for the process.

Once the components are placed into the paste, the paste must be cured to hold the components in place and to dry out the flux volatiles to prevent excessively quick outgassing which would cause mislocated components, disturbed solder joints, and the creation of solder balls. The drying cycle is relatively simple: elevate the temperature of the assembly to between 80°C (176°F) and 100°C (212°F) for 10-15 min. This can be accomplished using a standard type of recirculated hot air oven.

It can also be accomplished by using an in-line or conveyorized oven be it either infrared (Fig. 7.8), convection, or a combination of both. Infrared ovens are capable of doing both simultaneously: baking and curing the paste, activating the flux, and reflowing the paste. There are advantages and disadvantages, as we will see later.

Do not confuse the drying cycle with the past activation cycle. They are performed at different temperatures and for different time cycles. Don't believe for a moment that once a paste or cream has been baked its flux constituents have been activated; in all probability they have not.

7.6.1 Activation of the Flux and Reflowing

Activation of the flux and reflowing can be accomplished by a variety of methods, two of the more popular being the infrared and vapor phase systems. One must be cognizant that with either system the main funtion is to supply heat. These systems do not

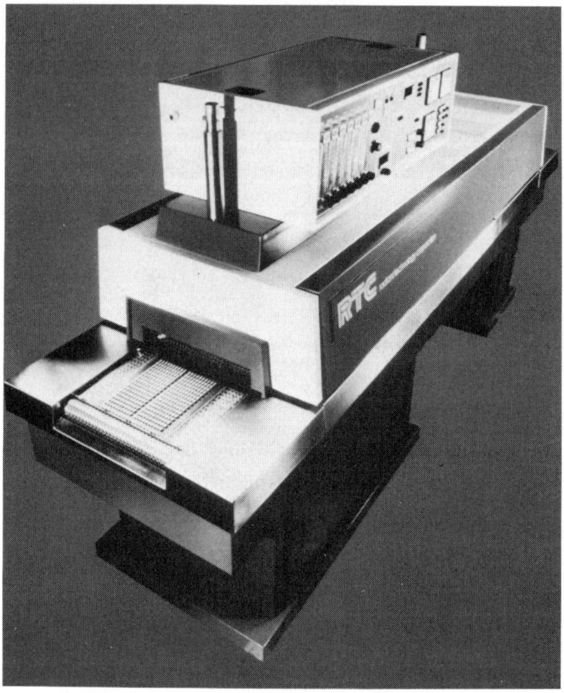

Figure 7.8 RTC Infrared system (slide). (Courtesy of RTC Inc., Cerritos, CA.)

create shorts, opens, or misaligned components. All those elements are created by the reflow process but are caused by other elements of the process such as improper paste thickness, component misalignment, paste misalignment, etc. This is why it was stated in the beginning that all elements must work together in surface mount assembly process.

Allowing for the fact that the infrared and vapor phase systems only supply heat, what are the differences? Infrared has a slower ramp-up time to reach soldering temperatures. The temperature profiles are usually modified to accommodate the prebake, flux activation, and reflow portions of the process.

7.7 INFRARED SYSTEMS

Long infrared wavelengths and diffused wavelengths are better than unfocused or focused infrared lamps because they are less color-sensitive and provide a more consistent heat across the product. Thermocouple the components and board to determine the temperature delta across the assembly and components. Through experimentation it has been demonstrated that board temperatures are higher than component temperatures during the reflow process in the infrared system.

Since the heat being generated by the infrared system is on both sides of the product, it does stand to reason that the printed wiring board will get hotter than the components. However, one must realize that although the mechanism which supplies the majority of the heat is convection, radiation still exists in infrared systems. As the printed wiring board heats up to the glass transition (Tg) temperature, the temperature at which the epoxy material changes from solid form to plastic form, the emissivity of the laminate will change and could be more susceptible to radiation abosrption. This would also cause the module to overheat and leave the components at a lower temperature.

Through industrial experimentation it has also been demonstrated that smaller components have a tendency to track the temperature of the laminate very closely, while larger components tend to remain cooler. Although this can be monitored through thermocoupling, it must be performed for each product being fabricated. One cannot rest on the belief that similar boards will behave similarly once submitted to identical conditions of the infrared system.

An example of some potential problem areas are the maximum and minimum temperatures necessary to create a quality assembly without degrading the base material or the components. The author suggests establishing upper and lower temperature limits for any product being fabricated. These limits are simply deduced by the end requirements:

- Don't destroy the board by overheating.

○ Reflow the solder joint.

Although they are simple, two basic temperatures are then implied.

○ The high end temperature should not be greater than 250°C (482°F) for a time span not to exceed 1-2 min.

○ The low end temperature must be above the melting point of the alloy of the solder paste. For 63-37 tin-lead alloy, this temperature is 183°C (361°F).

The problems start when the printed wiring board is populated with a wide variety of component sizes, large and small. The process temperature differential could become substantial; and, in fact, the high end of 250°C (482°F) could be exceeded and still have solder joints that are not created or reflowed.

Does this happen often? No. How can it be avoided? Thermocouple every board including component bodies and solder joints and know the actual temperatures to which the products are exposed. Prior to processing product, certify and guarantee the correct profile with the thermocouple board.

The best application for infrared reflow, from this author's perspective, is when all the components are similar in sizes and shapes. This simplifies the large temperature gradient and reduces the potential of process induced defects. It also minimizes the effect of board sizes to allow for even heating over the entire surface area of the board.

One segment still to be mentioned is the capability of the components to withstand the process temperatures. Close working relationships with component and design engineering groups are a must so that everyone is knowledgeable of both the materials and the process. This will help in reducing and eliminating process-induced defects (see Fig. 7.9).

Experiments must also be conducted to review and define the allowable amount of voids and whether or not they do impact the strength and reliability of the solder joint.

Figure 7.9 Vitronics SMD 700 series system. (Courtesy of Vitronics Inc., Newmarket, NH.)

7.8 VAPOR PHASE REFLOW SYSTEMS

Another reflow method that is generally accepted as the standard reflow process for surface mount is vapor phase (see Fig. 7.10). Vapor phase systems and processes are relatively simple when separated into its basic elements:

- Stainless steel containers capable of heating liquid material to a maximum temperature of 490-500°F (254-260°C)
- Liquid medium which boils at elevated temperatures

Put the two together and the heat source is created for the reflow process. Simple is the theory, complex is the application.

The heat transfer medium in vapor phase is the vapor blanket (see Fig. 7.11). As the product enters the vapor blanket, the fluorinert material immediately begins to condense on all the surfaces of the product. When the condensation stops, the product has reached the maximum temperature or the equilibrium temperature of the system.

A major advantage of the vapor phase system is the maximum temperature which the product will see or be exposed to during the reflow process. Also, as compared to the infrared reflow the minimum temperature exposure of the product will be higher, certifying that all the solder joints have been exposed to reflow temperatures.

Vapor phase unit systems can be both in-line or batch-type systems. In-line units are more conducive to high-volume production than batch units; however, they both have peculiarities which must be understood before the unit and processes can be used in manufacuring. Although this text mentions these, it cannot do justice to all the intricacies of the materials and byproducts evolving from a vapor phase process. The author highly recommends consultation with plant safety, industrial hygienist, nurse, and training personnel.

A typical fluorinert material used is Fc 70 [1]. This material has a boiling point of 419°F (215°C). The design of the

Figure 7.10 Dynapert-HTC IL-18CC. (Courtesy of Dynapert-HTC, Concord, MA.)

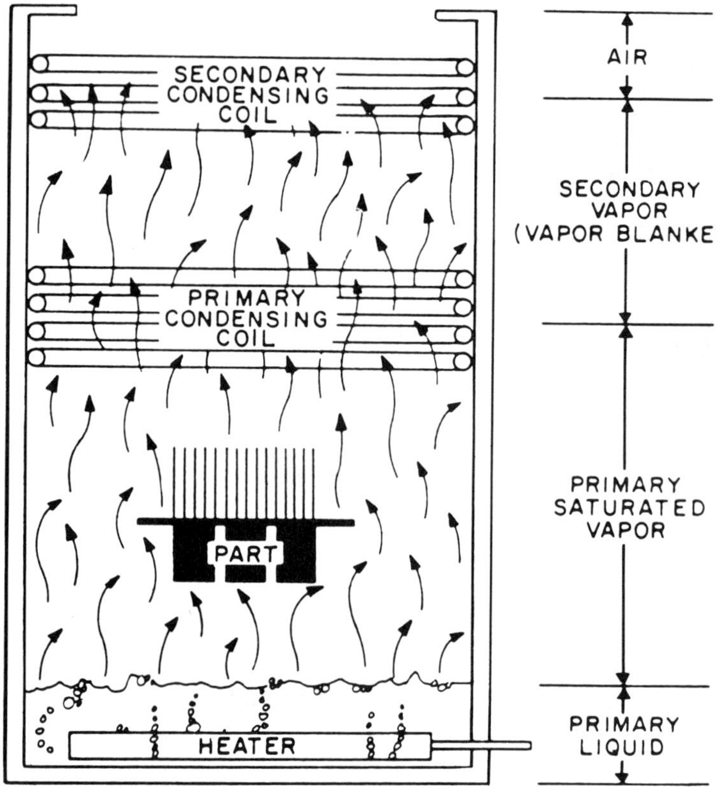

Figure 7.11 Dynapert-HTC sketch diagram (slide). (Courtesy of Dynapert-HTC, Concord, MA.)

process must be carefully planned to assure a good working environment. Equipment must be sized according to volume and anticipated product forecast.

A system generates a given quantity of caloric energy, and this energy is controlled through the temperature of the cooling coils. The cooling coils must be sized to remove the maximum caloric capacity of the vapor phase chamber. As product is introduced into th vapor zone, it also behaves as a mechanism to remove caloric energy for the system. This causes a variation or deviation from what is being removed by the cooling coils, hence as energy is being removed by the product the coils must be designed so they can vary to maintain equilibrium in the system. Pushing the capability of the system by introducing too much mass into the vapors will destroy the equilibrium of the system. This will eliminate the vapor blanket and impact the ability of the system to create enough heat for the creation of good solder joints. A system must not be overloaded. This implies that the caloric content of the vapor blanket is finite in volume; and although it is self-generating, its depletion can occur, and the system must be driven harder to maintain its equilibrium. This equilibrium is a fine balance; and if the systems are not properly sized, they will not be able to generate the necessary vapors to reflow the solder joints.

In planning for a system, the size and mass of the product plays an important role in sizing the equipment. A backplane loaded with connectors and solder performs will absorb more energy from the vapor blankets than a small 5 in. × 10 in. printed wiring assembly loaded with only surface mount components.

One of the cost algorithms is the material cost amortized over a total quantity of solder joints created. This creates a common denominator for comparison purposes of various processes. With this method, comparison to various heating methods and soldering processes are available for process justifications.

7.8.1 Safety

The thermal decomposition of Fc 70, perfluoroisobutelyne (PFIB), has been identified as a highly toxic material and difficult to capture and measure. The second most detrimental effect or vapor phase is the introduction of moisture into the system and the generation of hydrofluoric (HF) acid. This can be handled quite easily with acid separators or water separators.

7.8.2 Operational Usage of Vapor Phase

Again as with the infrared system, each product processed must be thermocoupled and profiled through the system. One of the characteristics of the vapor phase reflow system is the quick rise in temperature once the product contacts the vapors. It must be assured by component and design engineering that this temperature rise will not be detrimental to the components themselves. Another aspect of this quick temperature rise is the impact on moisture that may be entrapped within the devices. Quick vaporization may and in all probability will destroy the components due to the excessive pressure built up within the components. Hence, proper handling of components and storage must be carefully evaluated to certify that no degradation of components is occurring.

The dwell time in the vapor phase system is approximately 60-90 sec. During this dwell, many metal interactions are occurring, e.g., intermetallic growth, diffusion of plated surfaces, dissolution of like materials, and oxidation of surfaces. All these elements must be considered when designing and creating solder joints.

At all times we must remember that the single most important function of these heating systems is to generate heat to reflow the solder paste and not damage and printed wiring board. The long dwell time at elevated temperatures increases the growth of a Cu-Sn intermetallic which could be harmful to the reliability of the solder joint. Microsections of the joint should be made to verify the integrity of the solder joint and process paramaters. It is better to have a destructive test program in place to verify process parameters than use simple visual inspection. Visual inspection only identifies cosmetic criteria and does not allow an understanding of the intermetallics and grain structure of the solder joint.

7.9 SURFACE MOUNT PROCESS ANOMALIES

Surface mount soldering brings many process and product disciplines into play. The paste is first applied to some predetermined location and thickness. The component whose lead platings have been qualified is positioned into the paste, the paste is dried, and the assembly is submitted for reflowing of the solder paste to create solder joints.

Reiterating the procedural sequence, if all goes well, the solder wets both the component lead and the land area and visual inspection reveals no solder balls or other solder-type defects. This is not always the result, however, and an understanding of each step of the process is necessary to properly troubleshoot the defects and determine the cause of the problems.

Listed are some of the typical defects and causes of defects found after the reflow operation.

7.10 INSPECTION OF SURFACE MOUNT ASSEMBLIES

The inspection of surface mount assemblies will necessitate more detailed information and tooling to competently identify product goodness. Component densities, hidden solder joints, operator eye fatigue are all elements that will have to be addressed in developing an inspection process.

If the product is mixed mount technology, the crossover between plated through-hole quality requirements and surface mount requirements will have to be addressed. The solder joints look different and are quite difficult to see due to product densities and joint locations. The process must also be finely tuned with engineering and reliability data to certify that nonfunctional touchups and repairs are not performed. Rejections must be made on the assessment that the product will fail its form, fit, and function purposes if accepted. This is a major change from the behavoirial practices of through-hole technology of repairing for the sake of cosmetics.

There are many standards available through organizations such as the IPC and military specifications which identify acceptable and rejectable parameters for all types of SMT devices. Depending on the strength of the organization and testing capability that is accessible, these standards can be used as they are or modified to particular needs as the product requirements dictate. Contractual agreement with the military negate any changes or variation from their specifications without prior approval and justification. Therefore, involvement in organizations that are influential in the generation of the specifications is the forum one must use to change the criteria.

7.10.1 Open Solder Joints or Missing Solder

Open solder joints are usually found on "J-bend" devices or PLCC (plastic leaded chip carriers). Although solder paste was properly deposited on the land and all the other joints appeared to have flowed properly, one or two solder joints are open or missing solder. Upon investigation, the leads in question have a buildup of solder up onto the lead away from the land area. The soldered joints do not exhibit this phenomenon. Best industrial estimates lead us to believe that the reason for this problem is the heat transfer mechanism of the vapor phase fluid to the product being soldered. One explanation: the product enters the vapor zone and condensation begins immediately on the component leads and the printed wiring boards. The component leads, having less mass, heat up first to the point of reaching soldering temperatures. When the component leads have reached equilibrium temperature of the fluid, they create a path for the solder paste to flow, removing it from the land area. Once this temperature has been reached by the leads, it melts the solder paste and reflows the paste. The paste having nowhere to go climbs up the component lead leaving the pad void of solder, and hence creating an open joint.

This phenomenon does not occur as frequently when using the infrared reflow process. It is believed to be related to the speed at which the product is brought to soldering temperatures. Infrared, having a longer temperature/time rise, gives the flux more time to work on the oxides on both the pads and component leads. Hence, when the reflow temperature is reached, the parts to be soldered have also reached the reflow temperatures required to keep the solder joint intact.

This is the present estimation. Verifying the concept will require intricate instrumentation to individually profile the temperature rise of the component lead, the solder paste, and the land area. Until they are made homogeneous by the reflow process, they are individual relatives if you will, with unsimilar alloys, plating thickness, and surface contamination.

7.10.2 Solder Shorts

Solder shorts occur because of heavy paste deposits, misaligned paste deposition, or misaligned components. Experiments must be conducted to determine the correct process parameters and placement accuracies necessary to prevent all types of shorts.

7.10.3 Voids

Another element to consider is the potential formation of voids in the solder joint. Two types of voids are possible with solder paste: "built-in voids" and those induced after reflow due to thermal cycling [1]. As indicated in the referenced article, voids can be attributed to organics in the paste. Organics is the term used to describe the materials in the paste which are not metallic, such as the flux and oxides which may build up on the solder spheres of the solder. These are defined within the paste specification which identifies the quantity and type allowed for the paste to properly function. Once the amount of organics is known, the process is established to bake, remove, or volatize off those built-in organics.

7.10.4 Poor Solderability

When thin tin-lead coatings are present, the natural copper-tin intermetallic formation removes the tin from either the tin plating or the tin-lead plating. In the case of the tin plating, the tin combines with the copper to form a CuSn and CuSn intermetallic. The tin can be completely used in the formation of intermetallic, thereby leaving the entire surface unsolderable. Fluxing action will not remove or prepare the intermetallic surface for soldering. Another problem could be if outer tin-lead plating is thin; after the copper-tin intermetallic has formed, the remaining surface material is lead, which is not soluble in solder and has a reflow temperature of 620°F (326°C) which is above normal soldering temperatures.

7.10.5 Tools for Inspection

For solder joint inspection, magnification must be allowed. This magnification should be 3X at a minimum with stereomicroscopes as the overall tool for disputed defects. Special microscopes are available which allow the operator to view the solder joint from an angle without tilting the product. This has advantages over the monocular microscope where the product must be manipulated into position for accurate determination of the problem.

Electrical testers checking for shorts and opens are most beneficial and should be mandatory for all surface mount processes. Open solder joints are difficult to find and identify, and an electrical test can pinpoint the problem immediately. Solder shorts fall into the same category; let the tester identify the problem and location.

The inspection sequence in the assembly process should be as follows:

- ELECTRICAL TEST
  - Pass (meets electrical specifications)
    - Visual inspection
- BAD (fails electrical test)
  - Repair the defects
    - Repeat electrical test

The visual inspection should check for defects that will cause catastrophic failures in the field, i.e., insufficient solder that will not withstand the rigors of the process, misalignment where the joints are not mechanically secure, contamination, wrong parts, or missing parts that were not detected by the electrical test.

7.11 PLANT IMPLEMENTATION CHECKLIST

The following list identifies the steps which should be taken to properly implement a surface mount program in a corporation. The decision to implement surface mount is more than the ability to manufacture but a commitment from upper management to support the program, which will put many restrictions on past practices and procedures.

7.11.1 Operation Management Interest

- Is the operations staff supportive of introducing surface mount into manufacturing?
- Is the plant staff cognizant of the manufacturing intricacies of surface mount technologies?
- Do the operational goals of the corporation include surface mount types of products?
- Have specific surface mount products been identified?
- Is it cost-effective to manufacture these products with surface mount components?
- Have the financial goals for the next few years included surface mount development and capital equipment?
- Has a program manager been defined to oversee the entire surface mount operation?
- Is the operations staff aware of the intensive training necessary to properly implement surface mount manufacturing?

7.11.2 Project Team Assignment

- Identify project leader.
- Identify team members.
    - Manufacturing engineer

- ○ Quality engineer
- ○ Operations representative
- ○ Occupational health representative
- ○ Purchasing
- ○ Plant facilities engineer
- ○ Training coordinator
  - ○ Production control representative
  - ○ Test engineer
  - ○ Personnel representative
  - ○ Design engineering representative

○ Publish list of team members.

○ Create an implementation plan for surface mount. Include certification plan of process.

7.11.3 Project Plan and Schedule

○ Create project plan.

○ Publish plan and schedule.

○ Establish and schedule status meetings.

7.11.4 Training Development

○ Identify training needs.

- ○ Skills
- ○ Knowledge

○ Schedule training program.

7.11.5 Environmental Protection

○ Contact local environmental agencies for specifications.

- Identify hazardous material storage and waste facilities.
- Create an environmental safety plan for the facility.
- Certify proper ventilation within the surface mount area.
- Identify hazards of handling solder paste, i.e., lead ingestion.

7.11.6 Facilities

- Identify power, air, cooling, ventilation, drainage, disposal requirements for surface mount process.
- Prepare and schedule installation of equipment.
- Identify nonenvironmental safety requirements.
- Perform equipment maintenance.

7.11.7 Health and Safety Concerns

- Perform facility review.
- Perform equipment review.
- Create employee protection plan.
- Review material storage and handling plans.

7.11.8 Electrical Test

- Identify test requirements.

7.11.9 Quality Engineering

- Identify and implement training programs for responsible groups.
- Perform qualification experiments.
- Create and implement quality audits and inspection procedures.

- Certify that process documentation conforms to process.
- Assure that quality and reliability manufacturing support systems are in place to support the surface mount process.

7.11.10 Manufacturing Engineering

- Identify process capacity requirements.
- Define area layout.
- Identify correct equipment.
- Certify availability of documentation.
- Identify peripheral equipment needs.
- Identify manual assembly needs.
- Define rework and repair process including methods and procedures.
- Define necessary electrostatic discharge requirements.
- Identify process performance requirements.
- Identify material-handling requirements.
- Perform process capability studies.
- Define and establish the necessary process controls.
- Create documentation for all the equipment:
    - Operation
    - Maintenance
    - Calibration
    - Safety
    - Spare parts
    - Equipment acceptance

7.11.11 Production Control (Materials Control)

- Does present material tracking process accommodate the needs of surface mount process?
- Is there a business plan for the purchase of each component being bought?
- Does a detail specification exist for each surface mount part?
- Has each part been qualified to certify compliance to specifications?
- Is there an incoming inspection plan for all surface mount parts?
- Is there a plant to implement a vendor quality program?

These elements should suggest enough questions to properly address the introduction of surface mount into any manufacturing process. Additionally, it can also be used for the planning and development of a new surface mount assembly process.

ASSOCIATED READINGS

1. Donald H. Daebler, *Flux Dam Aids Cleaning Under Chip Capacitors*, Electronic Packaging and Prodction, September 1981, pp 131-136.
2. Robert E. Peterson, Jr. *Components and Hardware*, EDN, December 1982, pp 240-248.
3. W. L. Brodsky, F. D. Parker, *Development of a 68-Pin Multiple In-Line Package*, Electronic Packaging and Production, April 1981, pp 124-136.
4. Howard W. Markstein, *Computer Packaging: More Capability in Less Volume*, Electronic Packaging and Production, January 1981, pp 65-72.
5. Philip Marcoux, *Automated Surface Mounted Assemblies*, Circuits Manufacturing, June 1982, pp 62-65.

6. Donald Elliot, *Wave Soldering of Discrete Surface Mounted Devices*, Electri-Onics, September 1984, pp 41-44.

7. William M. Hastie, *Wave-Soldering Surface Mount Devices*, Circuits Manufacturing, January 1984, pp 48-52.

8. John E. Hale, Warren R. Steinacker, *Complete Cleaning of Surface Mount Assemblies*, Freon Product Division, date unknown.

9. Bruce Murray, *Choosing the Right Surface-Mount Adhesive*, Machine Design, October 1984, pp 89-93.

10. Dr. Wallace Rubin, *Smoothing the Way for Solder Creams*, Circuits Manufacturing, October 1983, pp 74-80.

11. Michael P. Cassidy, Bruce C. Fichter, Anthony G. Lubowe, *Trends in Western Electric Circuit Pack Manufacture*, PC Fab, January 1982, pp 26-37.

12. Emory Garth, Foster Fray, *Using Plastic Leaded Chip Carriers*, Circuits Manufacturing, March 1984, pp 42-50.

13. David Schoenthaler, *Soldering Surface Mounted Chip Carriers to Printed Circuits*, Part 1, Electri-Onics, January 1984, pp 25-28.

14. David Schoenthaler, *Soldering Surface Mounted Chip Carriers to Printed Circuits*, Part 2, Electri-Onics, March 1984, pp 15-17.

15. Noel C. Peterson, *Using Screened Paste Solder*, Circuits Manufacturing, June 1982, pp 60-61.

16. Joe Keller, *Super Solder Saver Suggestions*, Assembly Engineering, July 1982.

17. Dr. Armin Rahn, *Soldering of Leadless Components*, Electri-Onics, January 1983, pp 13-14.

18. Francis J. Dance, *Part 1: Can PWBs Meet the Surface Mount Challenge?* Electri-Onics, May 1983, pp 25-31.

19. Francis J. Dance, *Part 2: Can PWBs Meet the Surface Mount Challenge?* Electri-Onics, June 1983, pp 35-38.

20. Ralph Woodgate, *How to Wave Solder Discrete Chip Components*, Electri-Onics, June 1983, pp 21-22.

21. Dr. Ronald P. Anjard, Sr. *Solder Paste Compositions and Application Techniques for PWB's Part 1*, Electri-Onics, September 1983, pp 37-40.

22. John S. Carter, *Solder Paste Equals Efficiency–An Important Productivity Equation*, Insulation/Circuits, June 1978, pp 23-24.

23. *Mounting of Chip Carriers To PCB's and Ceramic Substrates Using Vapor Phase Soldering*, Production Soldering Clinic, Insulation/Circuits, March 1981, pp 27-28.

24. William H. Down, *Reliable Wave Soldering of Chip Components*, Electronic Packaging and Production, July 1981, pp 171-173.

25. Donald Ford, *Vapor Phase Soldering*, Circuits Manufacturing, February 1985, pp 97-126.

26. D. J. Peck, *In-Line Vapor Phase Soldering: An Economic Reality*, Electronic Packaging and Production, November 1982, pp 62-67.

27. Peter H. Moy, David Hallowell, Amalendu Sanyal, *In-Line Reflow Soldering: Not Just A Passing Vapor Phase*, Circuits Manufacturing, September 1983, pp 32-36.

28. Howard W. Markstein, *Vapor Phase Soldering*, Electronic Packaging and Production, July 1982, pp 33-44.

29. Jacques Barranger, *Liquid-Vapor Interfacial Tension: A Prime Factor in Solderability Measurements*, Electronic Packaging and Production, July 1980, pp 204-211.

30. Robert A. Reynolds, *SMDs Invade Military and Commercial Equipment*, Electronic Packaging and Production, February 1985, pp 130-133.

31. Tom Dixon, *SMT Calls for a Re-examination of Soldering Techniques*, Electronic Packaging and Production, February 1985, pp 96-103.

32. Ronald Pound, *Conductive Epoxy Is Tested for SMT Solder Replacement*, Electronic Packaging and Production, February 1985, pp 86-90.

33. Michael L Martel, *Curing and Solderability Methods in PTF and SM Attachment–Part 1*, Electri-Onics, November 1985, pp 37-39.

34. John P. Crisanti, Bhasker C. Desai, *Clip-On Terminals Solve CCC Connection Problems*, Electri-Onics, July 1984, pp 21-23.

35. Earl F. Lish, *Consideration for Laser Soldering Surface Mounted Connectors*, Electri-Onics, June 1984, pp 28-32.

36. Donald Ford, *A Common Sense Approach to Selecting a Solder Paste*, Circuits Manufacturing, August 1985, pp 60-64.

37. Howard W. Markstein, *Low TCE Metals and Fibers Prove Viable for SMT Substrates*, Electronic Packaging and Production, Janury 1985, pp 52-59.

38. Tom Dixon, *Unresolved Issues Hinder Full Realization of SMT's Benefits*, Electronic Packaging and Production, January 1985, pp 40-46.

39. Donald Utz, *Better Solder Printing with Stencils?* Circuits Manufacturing, October 1985, pp 43-46.

40. P. R. Jones, *Solder Joint Integrity Is Key to Temperature Cycling Resistance*, Electronic Packaging and Production, January 1985, pp 84-90.

41. Michael L. Martel, *Curing and Solderability Methods in PTF and SM Attachment–Part 2*, Electri-Onics, December 1985, pp 37-42.

8

Troubleshooting and Recommendations

8.1.22 Missing Solder–Fixture-Related
8.1.23 Missing Solder–Plated Through-Hole-Related
8.1.24 Missing Solder–Solder Mask-Related
8.1.25 No Foaming–Flux-Related
8.1.26 Nonwetting–All Conditions
8.1.27 Poor Soldering–Flux-Related
8.1.28 Shallow Fillet–Conveyor-Related
8.1.29 Shallow Fillet–Dirty Holes
8.1.30 Shallow Fillet–Excessive Oil in the Wave
8.1.31 Shallow Fillet–Flux Application-Related
8.1.32 Shallow Fillet–Heat Sink-Related
8.1.33 Shallow Fillet–Inner Layer Heat Sinking
8.1.34 Shallow Fillet–Lead to Hole Size Ratio
8.1.35 Shallow Fillet–Orientation-Related
8.1.36 Shallow Fillet–Preheat-Related
8.1.37 Shallow Fillet-Trailing Edge
8.1.38 Shorts–Lead to Etch
8.1.39 Shorts–Lead to Etch–Process-Related
8.1.40 Shorts–Lead to Lead–Design-Related
8.1.41 Shorts–Lead to Lead–Process-Related
8.1.42 Solder Balls–Damaged Solder Mask
8.1.43 Solder Balls–Random
8.1.44 Solder Balls–Solder Paste-Related
8.1.45 Solder Balls–Squeeze Out
8.1.46 Solder Balls–Squeeze Up
8.1.47 Solder Paste–Moving Components
8.1.48 Solder Paste Deposition Process Problems
8.1.49 Voids–Blow Holes
8.1.50 Voids–Blow Holes
8.1.51 Webbing–Contamination Related
8.1.52 Webbing–Flux Related
8.1.53 Webbing–Hot Boards
8.1.54 Webbing–Laminate-Related
8.1.55 Webbing–Mask Curing-Related
8.1.56 Webbing–Missing Solder Mask
8.1.57 Webbing–Pot Temperature-Related
8.1.58 Webbing–Splash-Related
8.1.59 White Residue
8.1.60 White Residue–Component Leads
8.1.61 White Residue–On the Etch Lines
8.1.62 White Residue–On the Laminate

8.0 INTRODUCTION

This chapter identifies defects and some of the associated causes and cures in order to prevent their occurrence. Invariably, during the process of assembling and soldering components to printed wiring boards, defects are created. Preventing these defects from occurring should be a major goal of any corporation because of the cost associated with creating defective product. Defective product impacts reliability performance, customer satisfaction, and company profits. To be successful, quality production is a must.

Therefore, to properly attack soldering problems, one has to identify and perform defect analysis work on the problems in order to isolate the failure mechanism. Defect analysis involves performing many laboratory functions on the defective parts, some of which are the following:

- Metallography study to review grain structures and crack propagation, and intermetallic growth
- Scanning electron microscope to perform intricate surface analysis
- X-ray diffraction for elemental metallic mapping
- Auger analysis to perform surface analysis
- Microscopes for high-power visual examination
- Thermal cycling to replicate fatigue failures
- Thermal shock to replicate product usage
- Temperature humidity conditioning to simulate corrosive growth from contaminants on surfaces

8.1 PROBLEM IDENTIFICATION AND RECOMMENDED TREATMENTS

The following list identifies some of the problems which occur as a result of the soldering operation, including the problems associated with the equipment.

8.1.1 Cleaning—Rosin-Related

Rosin is evident on the surface of the printed wiring assembly (pwa).

Explanation:

1. Improper top side pressure will result in nonwashing, leaving behind flux residues.
2. Low detergent concentration will prevent proper cleaning.

Recommended Treatment:

1. Check the operational procedure of the cleaning system to certify proper manifold pressure.
2. Check detergent concentration.

8.1.2 Component Preparation—Long Leads

Preforming and cutting operation is not correct.

Explanation:

Although the lead length may not interfere with overall component lead protrusion on the solder side of the pwb, it may create shorts with nearby metallic conductors.

Recommended Treatment:

Check equipment because lead length is too long.

8.1.3 Contaminated Flux

The flux is contaminated.

Explanation:

Contaminated flux can occur from a variety of reasons:

1. Oil dripping off the conveyor into the flux pot.

2. Water being brought into the fluxer through the air line.
3. Contaminated air stone from not being cleaned properly.
4. Dirt and particulate matter dropping into the fluxer from the modules being processed through the system.
5. Silicon contamination from the excessive use of silicon on power components.

Recommended Treatment:

1. Drain the flux pot and clean the brushes and air stone.
2. Wipe down the fluxer chimney and pot.
3. Replenish the fluxer with clean flux.

8.1.4 Contaminated Flux—Oil in the Flux Pot

The flux is contaminated.

Explanation:

The oil dripping off the conveyor has been dripping into the flux pot contaminating the flux. The flux looks grayish when this occurs. If the flux does not appear amber or clear in color, then it must be replaced.

Recommended Treatment:

1. Drain the flux pot and clean the brushes and air stone.
2. Wipe down the fluxer chimney and pot.
3. Replenish the fluxer with clean flux.
4. Wipe down the conveyor to remove the oil coming from the solder pot.
5. Make sure the drip plate is cleaned on both the conveyor rails.

8.1.5 Contaminated Flux—Poor Foam Head

The flux is contaminated.

Explanation:

The clogged stone was caused by allowing the stone to air-dry. The solvent evaporated and left behind the flux material in the pores of the stone. It is important to clean the stone daily in flux thinner. If the stone is completely clogged it will have to be soaked overnight to soften the flux residues and allow it to be removed.

Recommended Treatment:

1. Drain the flux pot and clean the brushes and air stone.
2. Wipe down the fluxer chimney and pot.
3. Replenish the fluxer with clean flux.

8.1.6 Contaminted Flux Filter—Water in Fluxer

The flux is contaminated.

Explanation:

The filter was not drained and water was introduced to the fluxer through the air line. This moisture will not evaporate going over the preheaters and will volatilize going over the wave. This will cause spattering of the flux which in turn cause voids and pinholes.

Recommended Treatment:

1. Drain the flux pot and clean the brushes and air stone.
2. Wipe down the fluxer chimney and pot.
3. Replenish the fluxer with clean flux.
4. Dry the air stone completely.

8.1.7 Conveyor Angle Adjustment—Wrong Conveyor Angle

Conveyor angle needs adjustment.

Explanation:

Incline conveyors were designed to help the peel back action of the solder. If the angle is wrong, solder shorts or icicles may be created.

Recommended Treatment:

Adjust the conveyor angle so that it is set 7-8°.

8.1.8 Conveyor Fast

The conveyor speed *exceeds* specification.

Explanation:

The contact time of the module in the wave is too short.

Recommended Treatment:

Reduce the conveyor speed to the module specification.

8.1.9 Conveyor to Slow—Icicles and Shorts

The conveyor speed is *below* specification.

Explanation:

The conveyor speed is too slow and is increasing the soldering time of the module in the wave. The added heat from the solder is burning off the flux from the soldering surface. As the module exits the wave, the surface tension between the module and the molten solder will not allow the solder to flow back into the solder pot, thus causing icicles and/or shorts.

Recommended Treatment:

1. Increase the conveyor speed to the module specification.
2. The depth of the module in the wave should be checked and adjusted so that the contact distance is 1-2 in. and the dwell time is 1-1.5 sec. Use a glass plate to measure this.

8.1.10 Dewetting

Major dewetting is occurring on both sides of the boards.

Explanation:

Dewetting occurring on both sides of the board implies poor preparation of the base material during board shop fabrication. This is proven by the inability of the wave solder process to reduce or eliminate the dewetting condition.

Recommended Treatment:

1. If major dewetting is evident on the pads and in some of the plated through holes, it can be argued that dewetting is also occurring in the plated through holes themselves and cannot be fixed. The boards must be scrapped.
2. If major dewetting is not impacting the soldering of the components to the boards, then the condition is acceptable and rework should not be performed.
3. Try to determine whether it is date code-related, vendor-identifiable, etc. This may lead to determination of the problem.

8.1.11 Dewetting—Minor Condition

Minor dewetting of the pads and plated through holes are preventing the smooth fillet transition of solder to the pad and barrel.

Explanation:

This problem can be caused by contaminated base copper material or by being exposed to high heat for an extended period of time, creating an increase in the intermetallic layer thickness. The intermetallic is not solderable; therefore the solder is no longer attracted by its own forces and pulls away from those locations.

Recommended Treatment:

1. If the dewetting is not greater than 25% of the circumference of the plated through hole, the board is acceptable. Do not touch it up.

2. If the dewetting is greater than 25% of th circumference, touch up/repair the plated through hole and accept the board. If it cannot be fixed, the board must be scrapped. Note: If the dewetting is due to poor fluxing, it can be fixed by refluxing and resoldering. If it is due to intermetallic, it cannot be fixed and the product should be scrapped.

8.1.12 Dewetting—Side Two

Dewetting is occurring on side 2 (solder side) on the etch lines.

Explanation:

This is caused by thick intermetallic layers preventing the solder from spreading evenly over the continuous area of the etch lines.

Recommended Treatment:

Provided the dewetting does not affect the actual wetting of the component lead to the plated through hole, the board is acceptable.

8.1.13 Dirty Oil

The oil is dirty and is creating sludge and dross buildup in the solder pot.

Explanation:

The oil is dirty if sludge can be seen in and around the sump area. This sludge is a dross material which will get pumped through the wave and deposited onto the surface of the joints and in the joints creating pinholes.

Recommended Treatment:

1. Drain the oil from the solder pot.
2. Clean off the sludge and dross buildup from the sump and within the area of the solder pot and the baffles.
3. Clean off the baffles.

4. Replace the oil with fresh oil and make sure the oil intake valve is covered approximately 1/8 in. of oil.

8.1.14 Dull Gray Plating Surface

The plating which is supposed to look shiny and silvery is dull and gray.

Explanation:

This may mean the boards missed the reflow or fusing operation. Electroplated tin-lead (63-37) has a matte finish appearance and when it is fused it changes to an alloy of solder (tin-lead) and appears shiny and smooth. If the alloy is high in lead, it will also be dull and gray, which is the way it is supposed to look.

Recommended Treatment:

Check the alloy. If it is 63-37 tin-lead, then the fusing probably has not been performed. Check the lot and it it is evident on the entire lot, return it to the fabricator.

8.1.15 Flux—Air Stone Plugged

The air stone is plugged solid with dried flux residues.

Explanation:

The flux in the stone hardens when allowed to dry. Therefore, if it is allowed to dry, it will plug the pores of the stone and will not allow the air to flow through, preventing the formation of a good head of foam. Also, as the flux softens from trying to push fresh flux through the stone, the old flux will contaminate the flux pot. It is important to clean the stone daily.

Recommended Treatment:

It must be soaked overnight (minimum 16 hr) in flux thinner to soften and remove the flux residues.

8.1.16 Icicles—Contaminated Solder

The solder pot has become contaminated and sluggish, creating shorts.

Explanation:

Contaminated solder, which is contaminated with other metallic elements becomes sluggish. This sluggishness prevents the solder from peeling away quickly from the bottom side of the assembly during the soldering operation. When this occurs, icicles are usually formed.

Recommended Treatment:

Change solder in solder pot. It has become contaminted and sluggish therefore creating icicles on the tips of the leads.

8.1.17 Icicles—Side Bar Adjustment

When the side plate adjustment is too low, the solder will form an icicle on the tip of the lead.

Explanation:

This allows the solder to fall away much quicker due to the speed of the wave falling down back into the solder pot. This may cause icicles on the tip of the leads.

Recommended Treatment:

Adjust side plates to be no more than 1/8 in. below the edge of the solder pot.

8.1.18 Icicles—Warpage-Related

Icicles are evident on the trailing edge of the pwb as it exits the wave solder system.

Explanation:

1. The board could be too tight in the finger conveyor which would cause excessive warpage creating the icicles.
2. Large metallic surfaces on the trailing edge not allowing the solder to drain properly off the pwb as it exits the wave solder pot.

Recommended Treatment:

1. Remove icicles with a solder iron.
2. Generate ECO to remove the large metallic surface from the edge of the pwb, or cover it with solder mask.
3. Convert the product to solder mask over bare copper.

8.1.19 Long Leads

The leads are too long and violate the requirements of side 2 or solder side protrusion.

Explanation:

The leads are longer than allowed per the specification. The potential exists that theleads will short to the adjacent board or chassis during actual usage, or if it is bent over it could short to adjacent leads or pads.

Recommended Treatment:

Trim the leads to the product requirements.

8.1.20 Missing Solder—Conveyor Related

Missing solder is occurring near the edge of the pwa where the board sits in the fingers.

Explanation:

The fingers are obstructing the solder from contacting the hole in which case they are either bent or the wrong type of finger. Or the holes may be so close to the edge that the conveyor fingers are acting as heat sinks preventing the holes from filling.

Recommended Treatment:

1. If conveyor fingers are bent or the wrong type, replace or fix the fingers.
2. If holes are too close to the edge, fixtures must be designed to help the soldering and prevent this problem.

8.1.21 Missing Solder—Conveyor-Related

The pwa is not sitting properly on the conveyor fingers.

Explanation:

It is not properly seated in the fingers, so that as it goes over the wave, when it heats up it bends up and in fact never contacts the wave at all, creating an entire area which is not soldered. The conveyor could also be out of adjustment and not parallel to the surface of the wave. When this occurs, one side of the pwb will solder and the other will not.

Recommended Treatment:

1. Reseat the board in the conveyor and resolder. This will eliminate the missing solder.
2. If the conveyor is not adjusted parallel to the wave it must be fixed so the entire board contacts the wave equally across its entire width.

8.1.22 Missing Solder—Fixture-Related

The design of the fixture is preventing the holes from filling.

Explanation:

The holes on the edge of the pwa are too close to the fixture, preventing the solder from contacting the plated through holes to complete the soldering operation. By relieving the fixture around these particular holes would enhance the soldering of those components. The fixture could also be preventing the board from reaching soldering temperature by acting as a heat sink. This would prevent the hole from filling and could in fact appear as nonwetted solder joints.

Recommended Treatment:

Change the design of the fixture to relieve the board.

8.1.23 Missing Solder—Plated Through Hole-Related

Missing plating in the hole prevent the hole from filling with solder.

Explanation:

Missing plated through holes are the cause of poor plating practices in the boards shops. The boards could have missed a process step or been introduced to an improperly controlled plating bath. In any case, when the solder is missing from the hole, there is no path for the solder to wet the hole. This means that the solder will not be able to rise in the hole, causing an empty hole or a missing solder condition.

Recommended Treatment:

When this occurs the only remedy is to scrap the boards.

8.1.24 Missing Solder—Solder Mask-Related

Permanent solder mask in the plated through hole is preventing hole from filling with solder.

Explanation:

In some cases, permanent solder mask is applied over plated through holes (vias) or noncomponent holes to prevent hole filling to reduce and eliminate solder shorts. When this occurs, the nonfilling is acceptable.

Recommended Treatment:

1. If the permanent solder mask is over component holes, the solder mask artwork must be changed to open up the hole and free it from the solder mask.
2. If the solder mask is in the plated through holes; this was caused by poor solder mask application in the board fabrication process. Typically the entire board exhibits this condition and none of the holes will fill. If this exists, send it through the wave solder system a second time. If the holes still don't fill, the board must be scrapped.

8.1.25 No Foaming—Flux-Related

The flux-specific gravity is too *high.*

Explanation:

When the flux gets too thick, it will not foam in a foam fluxer system.

Recommended Treatment:

Add thinner until the specific gravity is within the tolerances recommended by the supplier.

8.1.26 Nonwetting—All Conditions

Nonwetting is the result of poor preparation and contamination of the base material.

Explanation:

This problem is caused by contamination of the base material. A typical contaminant is organic and not metallic; therfore it is not removed by fluxes and soldering processes.

Recommended Treatment:

Scrap the board. It cannot be repaired and will create a highly probable intermittent condition.

8.1.27 Poor Soldering—Flux-Related

The flux-specific gravity is too *low.*

Explanation:

When the flux-specific gravity is too low, it is not active enough to remove contaminants from the pads and leads to prepare these for soldering. The surfaces will exhibit poor solderability which will not allow the solder to wick up the leads and plated through holes, causing shorts, icicles, and dewetting defects.

Recommended Treatment:

1. Add more flux to bring the specific gravity to within recommendation of supplier.

2. Change the flux if needed.
3. The flux viscosity should be temperature-adjusted with a new control point set. If a foam fluxer is used, then the flux bath may need agitation to ensure an accurate reading of flux viscosity.

8.1.28 Shallow Fillet—Conveyor-Related

The conveyor speed is below specification.

Explanation:

The conveyor speed is too slow and is increasing the soldering time of the module in the wave. The added heat from the solder is keeping the solder molten longer and allowing it to drain out of the plated through hole, leaving a shallow fillet. This is more likely to occur in plated through hole with IC leads. This is due to the lead-to-hole size ratio. The lead is very small compared to the hole size; therefore the only thing holding the solder in the path is the cohesive forces of the solder, whereas if the lead were larger then the forces holding the solder in the hole would be adhesive forces, which are greater and would prevent the solder from falling out of the plated through hole.

Recommended Treatment:

Increase the conveyor speed to the module specification.

8.1.29 Shallow Fillet—Dirty Holes

The plated through holes are contaminated and will not fill with solder.

Explanation:

Contaminated holes with dust and loose matter will not be cleaned with the fluxing operation. The flux is designed to remove mild oxides, not particulate matter or fingerprints. When this condition occurs, the holes will not fill with solder.

Recommended Treatment:

1. Check all the boards and see if it is vendor or date code-oriented. Perhaps there is a bad lot of boards.
2. Check board storage. Improper storage or not having a file system will increase the amount of oxidation and contamination on the board.
3. The boards can be returned to the vendor for cleaning or try running them through the aqueous cleaner to see if that helps; if not, return them to the vendor. Do not try to use the boards in this condition or try to touch them up. They will have high failure rates in the field due to intermittent types of solder joints.

8.1.30 Shallow Fillet—Excessive Oil in the Wave

Shallow fillets are seen and the oil intermix in the wave is heavy.

Explanation:

When the oil intermix is too heavy because there are large streaks of oil evident on the surface of the wave. This extra oil enhances the mobility of the solder, thereby allowing it to flow out of the plated through holes. This creates shallow fillets in the majority of all the plated through holes.

Recommended Treatment:

1. Using a stainless steel socket, turn the oil valve down until the volume of oil in the wave is a medium intermix, where the streaks of oil disappear and medium size droplets are observed on the surface of the wave.
2. Check the side plate adjustment to certify no extra build-up of dross in the side plate areas. Check both the front and the back side plate areas.
3. If the side plate areas are readjusted, then the oil valve will need to be readjusted.

8.1.31 Shallow Fillet—Flux Application-Related

The board did not properly contact the flux in the fluxing operation.

Explanation:

For the solder to properly flow up through a plated through hole, proper fluxing is necessary. This means that the flux must contact the internal metal portion of the barrel to prepare the surface for soldering. Flux removes metallic oxides and prevent the oxidation of the surface. If this is not accomplished the solder will not be able to wet the surfaces of the barrel and lead if one is in the hole and improper hole filling will occur.

Recommended Treatment:

1. Turn the pressure up in the fluxer unit to raise the level of the foam head.
2. Check the brushes in the fluxer to verify that they are holding the foam head to the correct height.
3. The bubbles in the foam are too large and not contacting all the surfaces of the boards. Turn down the pressure in the air stone.

8.1.32 Shallow Fillet—Heat Sink-Related

Heat sink areas are preventing the holes form filling.

Explanation:

When the holes are near large heat sink areas, the process-operating parameters must be reviewed to certify the proper preheat temperature, conveyor speed, and solder pot temperature.

Recommended Treatment:

Generate an engineering change order (ECO) to create heat relief for the plated through holes.

8.1.33 Shallow Fillet—Inner Layer Heat Sinking

Inner layer tied to the plated through-hole barrel are causing shallow solder condition.

Explanation:

The inner layers of any printed wiring board are copper and serve the function of either carrying electrical signals or acting as ground or voltage planes. The implication is that they are copper and are of a certain thickness, which makes them act as heat sinks. This heat-sinking effect will prevent the normal heating of the plated through hole and cause the solder to solidify before reaching the top side of the board. To get proper hole filling, the entire plated through hole must reach soldering temperatures (183°C; 361°F). If it does not then the solder will solidify and the filling of the hole will not be accomplished. More preheat is necessary to overcome this heat-sinking effect of the inner layers. It is also important to understand the thickness of the inner layers, as this will also impact how the boards must be soldered.

Recommended Treatment:

1. Increase the preheat to overcome the heat sinking of the inner layers.
2. Slow down the conveyor speed to increase the heating time in the preheater section of the system.

8.1.34 Shallow Fillet—Lead to Hole Size Ratio

The finish component hole size is too large for the lead, allowing the solder to drain out of the plated through hole.

Explanation:

Large holes have the capacity to withhold a lot of solder. Additionally, if the component lead is small, that increases the amount of solder in the plated through hole. This excess solder is being held in the hole by its cohesive forces, i.e., the forces of holding itself together. These forces are not strong enough to overcome the forces of gravity pulling on the weight of the solder, hence the solder is drawn out of the hole or falls out of the hole. If the hole were smaller or the lead were larger, then the solder would be held in the hold by cohesive forces. These are the forces created by the wetting action of the solder to the surface areas in the hole, which in this case is the lead diameter and length and the surface area of the plated through hole. These forces are greater than the cohesive forces and the condition of shallow fillets would no longer exist.

Recommended Treatment:

1. Generate an ECO to reduce the hole size.
2. The component leads from a particular vendor may be too small in diameter, thus creating this condition. In that case there is nothing to be done and we will live with the condition. This situation is common when many vendors are utilized for particular components.

8.1.35 Shallow Fillet—Orientation-Related

The insufficient solder is on the leading edge of component leads.

Explanation:

The excess oil in the wave will enhance the wetting action of the solder in the pot and cause the solder to be pulled out of the plated through hole. Rotating the board 90° will change the orientation of the lead in relationship to the direction of the flow of the wave and prevent the solder from being pulled out of the plated through hole.

Recommended Treatment:

1. Reduce the oil intermix in the wave.
2. Rotate the board 90° in the conveyor.

8.1.36 Shallow Fillet—Preheat-Related

The module surface temperature on the component side is too *low*.

Explanation:

The preheat temperature is too low to activate the flux and prepare the surfaces for soldering. The solder will not properly wick up the plated through holes and form a good solder joint. Additionally, as the module exits the wave and if the conveyor speed is too fast, the solder will not flow back into the solder pot causing icicles and/or excess solder defects.

Recommended Treatment:

1. Increase the preheating to the module by increasing the preheat temperature.
2. Increase the preheating to the module by lowering the conveyor speed.
3. After you have made the adjustment, check to ensure that the solder is flowing up through *all* plated through holes. Check the preheater for equal head distribution along the entire width of the module. The heating elements may need calibration or replacement.

8.1.37 Shallow Fillet—Trailing Edge

The trailing edge of component leads are exhibiting insufficient solder.

Explanation:

When the trailing edge has shallow fillets it is due to the lead still being in the solder pot after the hole itself has exited the pot. This condition allows the lead to remain hot, thereby allowing the solder to run down the lead and emptying the plated through hole. By running the conveyor a little faster, less heat would be applied to the board and the solder would not have enough time to fall out of the plated through hole.

Recommended Treatment:

1. Turn the oil down in the wave.
2. Run the conveyor a little faster.

8.1.38 Shorts—Lead to Etch

The leads are clinched over too much and touching the adjacent etch lines.

Explanation:

The component leads are too long and touching the adjacent etch lines or component leads.

Recommended Treatment:

Check the insertion equipment to certify that the clinch angle is sufficient to hold the component in place but not short to adjacent etches or other components. If the components are manually inserted, the leads should not be clinched but trimmed to the correct length.

8.1.39 Shorts—Lead to Etch—Process-Related

The manual clinching of the component leads was done in the wrong direction.

Explanation:

Clinching the leads in the wrong direction could result in shorts between the adjacent component leads and nearby etch lines.

Recommended Treatment:

Leads should be clinched along the incoming etch line to the plated through hole.

8.1.40 Shorts—Lead to Lead—Design-Related

The design layout of the board is incorrect.

Explanation:

The components are oriented so that their leads are across nearby etch lines, creating shorts.

Recommended Treatment:

The design specifications should be reviewed and compared to the requirements of the product. The components should be oriented with their long axes parallel to the direction of travel going over the wave solder system. If it is impossible to reorient the module due to gold finger orientation, then either generate an ECO to correct the problem or live with it.

8.1.41 Shorts—Lead to Lead—Process-Related

The orientation of the board going over the wave is incorrect.

Explanation:

This allows the components to be oriented incorrectly, which creates solder flow problems during the wave soldering process. This hinders the smooth flow of solder away from the component leads and allows the solder to remain behind and create shorts.

Recommended Treatment:

Turn the board 90°, allowing the IC-type components to be traveling with their long axes parallel to the direction of travel in the wave solder system. If the gold fingers prevent this from occurring or if the board is too narrow, then the design must be reviewed and corrected.

8.1.42 Solder Balls—Damaged Solder Mask

The solder balls are adhering to the solder mask, but in locations where there was apparent damage to the mask. These locations could be around components which were automatically inserted or the leads were manually clinched over.

Explanation:

The solder mask was damaged during the insertion process and when the solder operation was performed, the solder on the etch lines protruded from those areas and created solder balls. It must be noted that the solder mask does not adhere to the tin-lead etch lines.

Recommended Treatment;

1. Check handling procedures to certify that no damage is occurring and damaging the solder mask.
2. Readjust the cut-and-clinch mechanism so it does not contact the bottom side of the board during the insertion process.
3. Convert solder mask over tin-lead boards to solder mask over bare copper technology.

8.1.43 Solder Balls—Random

The solder is sticking to the solder mask during the wave-soldering operation.

Explanation:

This can be caused by not enough oil in the wave, too much drying of the flux over the preheaters, or uncured solder mask. As the solder peels off the surface of the printed wiring board, it does not peel off smoothly and sticks to the surface of the mask itself. As the dry-film mask goes over the wave, it softens up and the solder from the wave sticks to it, leaving behind solder balls or a webbing type of solder balls.

Recommended Treatment:

1. Turn up or increase the quantity of oil in the wave.
2. Use a higher solid content flux. The flux helps the peeling action of the solder wave as the board exits the wave solder pot.
3. Cure the solder mask. Check with vendor requirements for proper curing parameters.

8.1.44 Solder Balls—Solder Paste-Related

After the reflow operation, solder balls are evident over the entire assembly.

Explanation:

Solder balls are the results of many elements:

1. Moisture in the solder paste.
2. Contamination on the surface of the pwa.
3. Temperature ramp-up in reflow operation is too quick and flux vehicle volatizes off, exploding the solder paste.
4. Improper baking of the solder paste prior to reflow.
5. Flux system has lost its activity due to excessive shelf life.

Recommended Treatment:

Keep the solder paste jars sealed until time for use. If refrigerated, remove from the refrigerator and allow the paste to reach room temperature before opening the jar. This will prevent the moisture from the environment from condensing within the jar.

Always buy fresh paste. Do not keep more than 2-3 weeks supply of paste as inventory. Certify that the vendor does not keep solder paste in his inventory for your use. Vendor must make solder paste fresh with every order. This will prevent the flux from working on the oxides of the solder powder while it is in the jar.

Certify that the printed wiring boards are clean before solder paste application. Make sure you handle the boards by the edges as opposed to the surface because with surface mount components any touching of the surface impacts the solderability of the individual land areas or pads.

Verify with the vendor the best solder paste for the individual solder process. The paste must be made specially for the individual process. The parameters which must be reviewed are the ramp-up temperatures of the preheat oven and the reflow chamber.

8.1.45 Solder Balls—Squeeze Out

The solder balls are squeezing out beneath the solder mask from a nearby etch line.

Explanation:

As the solder reflows on the etch lines beneath the solder maks, it expands, and once it finds a weak spot it squeezes out and forms tiny solder balls. This typically happens in areas where the plated through-hole clearance pad is misregistered and an adjacent etch line is not properly covered with the solder mask. Or the solder mask is not properly exposed as is the case with photo imageable solder mask and there is a small footprint or no footprint of the solder mask on the laminate. The footprint is the portion of the solder mask which adheres to the laminate near any etch line, it should be a minimum of 0.001 in. for wet film solder mask and 0.003 in. for dry film solder mask.

Recommended Treatment:

Check footprint of solder mask. This is the portion of the solder mask which is in contact with the laminate material between the etch line and the plated through hole.

8.1.46 Solder Balls-Squeeze Up

The solder balls are coming up through the solder mask from an etch line or pad beneath the solder mask.

Explanation:

This typically occurs with dry film solder mask which is too thin or is applied over an etch which is too thick. As the solder reflows beneath the solder mask, it expands. Since the mask is thin and brittle, cracking occurs creating weak spots in the solder mask. As this condition occurs, weak spots are created allowing the solder to escape generating little solder balls on the solder mask.

Recommended Treatment:

1. Use thicker mask to prevent solder mask cracking.
2. Check plating thickness on the traces or etch lines; many times overplating does occur which causes this problem.

8.1.47 Solder Paste–Moving Components

The components have moved in the solder paste.

Explanation:

The solder paste is usually baked to prevent this from occurring. Additionally, the placement of the components may be inconsistent, positioning the components incorrectly. If any components are touched prior to reflow and bake, they will be dislodged from theri original position.

Recommended Treatment:

Verify the bake temperature profiles to certify the solder paste is in fact drying and holding the components in place. Check the "Z"-axis placement of the placement equipment. Any warpage in the board will effect the location of the board and when components are placed they may in fact be either dropped or compressed into the pwa. In either case, they will not be placed accurately. If they are manually assembled, extreme care should be taken to prevent extra moving of the component. In fact this method is not recommended for high quantity of devices per assembly.

8.1.48 Solder Paste Deposition Process Problems

The solder paste is not screening or stenciling correctly.

Explanation:

Check the slump test for the solder paste. If the solder paste does not pass the slump test, it will not remain on the pads where it was deposited. If the paste does not pass through the stencil or screen, check the sphere size of the powder. The wrong sphere size will prevent the paste from being deposited. Also check the actual outline of the solder spheres, which must be round. If they are needlelike in structure, they will not be able to pass through a stencil or screen.

Recommended Treatment:

Verify in the specification the size of the sphere from which the solder paste must be fabricated. Also verify the quantity of powder in the final solder paste and include the flux type to be used in the system. This will guarantee a certain joint size when reflowed. Perform fineness of grind test to determine the actual sphere sizes. Check the stencil or screen for proper tension and proper snapoff to deposit a crisply defined pad of solder paste.

8.1.49 Voids—Blow Holes

Boards need baking.

Explanation:

Nonbaked boards are susceptible to the absorption of moisture. In fact, they behave as a sponge. When subjected to the heat of the solder pot, the moisture quickly vaporizes and turns to steam. The pressure created by this evaporation will cause the solder to blow out of the plated through holes, creating voids.

Recommended Treatment:

Bake the boards.

8.1.50 Voids—Blow Holes

Poor plating in the holes will cause serious blow hole problems.

Explanation:

Poor plating in the plated through hole, which could be caused by smearing of the drilling operation or improper cleaning of the hole before copper plating, will cause voids in the barrel of the plated through hole. This will cause the entrapment of plating solvents in the fibers of the laminate and these will volatilize during the soldering operation causing blow holes.

Recommended Treatment:

Scrap the module.

8.1.51 Webbing—Contamination-Related

Contamintion in the solder pot is causing the solder to act sluggishly thereby creating webbing shorts.

Explanation:

Contaminated solder has a different viscosity from pure solder and this increase in viscosity makes the solder sluggish and non-mobile. Hence, when the board exits the wave solder pot the solder does not peel off the bottom of the board and adheres to it, creating webbing shorts.

Recommended Treatment:

Measure the elemental solder contamination monthly to ensure that it is within the specified tolerances.

8.1.52 Webbing—Flux-Related

Webbing on the surface of the modules can be caused by the adhesion of the solder to the surface of the board being soldered.

Explanation:

Improper fluxing will create a nonslippery surface preventing the solder from peeling off. With this difficulty of peeling off, the solder adheres to the solder mask during the soldering operation. With the solder not being able to peel off the surface, webbing occurs. Poor flux head in the pot can have a variety of causes; however, the two most prevalent are:

1. Hot boards which would knock down the foam head
2. Silicon in the flux which prevents the boards from soldering.

Recommended Treatment:

Do not put hot boards onto a wave solder system which uses a foam fluxer.

8.1.53 Webbing—Hot Boards

Solder webbing is occurring across the surface of the board.

Explanation:

Foam flux is very sensitive to the foaming characteristics. Any heat applied to the foam will destroy the flux bubbles and also reduce the amount of flux being applied to the surface of the modules.

Recommended Treatment:

Check the board temperature before setting it onto the conveyor. Hot boards will destroy the flux head and prevent the proper fluxing of the boards.

8.1.54 Webbing—Laminate-Related

The webbing of the solder is not across the etch lines, but on the laminate itself.

Explanation:

Webbing is the result of the solder adhering to the laminate of the printed wiring board and shorting out adjacent etches. If the webbing is not across the etch lines, then perhaps solder mask has been used on the board and the webbing action is on the surface of the solder mask and can be addressed by further questioning.

Recommended Treatment:

1. Cure the laminate.
2. Apply solder mask to the boards.

8.1.56 Webbing—Mask Curing-Related

Solder webbing is occurring on the surfaces of the solder mask.

Explanation:

Improperly cured solder mask will not repel solder from the wave solder system easily. The solder will stick to the surface of the solder mask creating solder shorts or other metallic particulate matter. This particulate matter can be dislodged and create shorts throughout the assembly.

Recommended Treatment:

1. Check the solder mask for proper curing.
2. Cure the solder mask properly.

8.1.56 Webbing—Missing Solder Mask

The webbing is occurring on the laminate material and shorting across adjacent etches, creating electrical shorts.

Explanation:

The intent of solder mask is to prevent solder from adhering to the laminate material and shorting out to adjacent etch lines. Webbing can occur on the laminate if the laminate is not properly cured or the etch lines are oriented in the wrong direction for the soldering operation.

Recommended Treatment:

Apply solder mask to the surface of the modules.

8.1.57 Webbing—Pot Temperature-Related

The solder pot temperature is too low causing the solder to we web across the surface of the board.

Explanation:

Low solder temperature changes the viscosity of the solder and its ability to move around quickly. When this occurs, the solder does not peel off the bottom side of the module is quickly curing the soldering process. This causes the solder to stick to the board and create webbing shorts.

Recommended Treatment:

Raise the temperature of the solder pot to 495°F.

8.1.58 Webbing—Splash-Related

Splashes of solder from the solder pot are causing webbing shorts.

Explanation:

As the solder is pumped through the wave generator, it falls into the side plate area. The side plates should be adjusted so the solder does not splash. The height of solder in the side plate areas should be no more than 1/8 in. from the top of the solder pot.

Recommended Treatment:

Adjust the side plates of the solder pot to prevent the solder from falling too far and creating splashes.

8.1.59 White Residue

White residues are evident on the laminate material.

Explanation:

This condition occurs primarily due to a combination of a poor cleaning process and defective board material.

Recommended Treatment:

This can have a variety of causes, some of which can only be defined in the laboratory.

8.1.60 White Residue—Component Leads

There is a white residue on the component leads, not on the laminate material.

Explanation:

The rosin flux was partially removed and looks as though it were bleached. It can be flaked off with a sharp tool. This is cured by cleaning aqueous cleaner or solvent cleaner and recharging it with the proper concentration of detergent or new solvent. If the film is over the component leads, it is an oxide resultant on the exposed metal leads. This condition is caused by not cleaning the product soon enough after wave soldering. The flux reacts with the metal surfaces and this cannot be washed off and removed.

Recommended Treatment:

1. Check detergent concentration.
2. Check rinse section of cleaner system.
3. Clean the boards within 5 min of soldering, and keep the cleaning systems clean and properly maintained.

8.1.61 White Residue—On the Etch Lines

White residue on etch line and not on laminate material.

Explanation:

Contaminted solder will create grainy dull solder joints.

Recommended Treatment:

This is caused by contaminated solder. Check contamination level in solder pot.

8.1.62 White Residue—On the Laminate

White residue on the laminate material, side 1.

Explanation:

Concentration of contaminants is too high and being redeposited on the module.

Recommended Treatment:

Dirty cleaning solvent.

Index